Ceramics for Environmental Systems

Ceramics for Environmental Systems

Ceramic Transactions, Volume 257

A Collection of Papers Presented at CMCEE-11, June 14–19, 2015, Vancouver, BC, Canada

Edited by
Lianzhou Wang
Nobuhito Imanaka
Waltraud M. Kriven
Manabu Fukushima
Girish Kale

Volume Editors
Mrityunjay Singh
Tatsuki Ohji
Alexander Michaelis

The American Ceramic Society

WILEY

Published by John Wiley & Sons, Inc., Hoboken, New Jersey.
Published simultaneously in Canada.

Library of Congress Cataloging-in-Publication Data is available.

ISBN: 978-1-119-23444-9
ISSN: 1042-1122

10 9 8 7 6 5 4 3 2 1

Contents

Preface vii

PHOTOCATALYSTS FOR ENERGY AND ENVIRONMENTAL APPLICATIONS

Effect of Structural Properties on the Photoelectrochemical Performance of TiO_2 Films 3
A. K. Alves, A. C. Teloeken, F. A. Berutti, and C. P. Bergmann

Photocatalytic Degradation of Dyes using MWCNT-TiO_2 Composites as Catalyst 13
F. A. Berutti, A. P. Garcia, A. K. Alves, S. Da Dalt, and C. P. Bergmann

Synthesis of the TiO_2-Long Lasting Phosphor ($Sr_4Al_{14}O_{25}$:Eu^{2+},Dy^{3+}) Composite and Its Photocatalytic Reaction Properties 23
Jung-Sik Kim, Hyun-Je Sung, and Sang-Chul Jung

Development of Microtextured Titanium Dioxide Surface by using Microcutting Techniques 35
J. Shimizu, T. Yamamoto, L. Zhou, T. Onuki, and H. Ojima

Morphology Control and Photocatalytic Activity of TiO_2 Film 43
Jinshu Wang, Hongyi Li, Junshu Wu, Qian Cai, Yilong Yang, and Bingxin Zhao

ADVANCED FUNCTIONAL MATERIALS, DEVICES, AND SYSTEMS FOR THE ENVIRONMENT

Electrochemical Devices with Oxide Ion Electrolytes for Formation of Hydrogen and Decomposition of Carbon Dioxide from the CH_4–CO_2 Mixed Biogas 59
Yoshihiro Hirata, Soichiro Sameshima, and Taro Shimonosono

Gastight, Closed Pore Inclusive Porous Ceramics through a Superplastically Foaming Method 69
Akira Kishimoto, Atsuki Tohji, Takashi Teranishi, and Hidetaka Hayashi

Cyanosilylation of Benzaldehyde with Trimethylsilyl Cyanide Over A-Site Metal Substituted Perovskite-Type Oxide Catalyst Prepared by Thermal Decomposition of Heteronuclear Cyano Complex Precursors 81
Syuhei Yamaguchi, Hiroki Wada, Takahisa Okuwa, and Hidenori Yahiro

GEOPOLYMERS, INORGANIC POLYMER CERAMICS, AND SUSTAINABLE COMPOSITES

Nanoparticles Seeded Geopolymers 93
Matteo Pernechele, Tom Troczynski, and Marek Pawlik

NH_3BH_3 and $NaBH_4$ Enclosed in Geopolymers and Zeolites 105
C. H. Rüscher, L. Schomborg, Z. Assi, and J. C. Buhl

MACROPOROUS CERAMICS FOR ENVIRONMENTAL AND ENERGY APPLICATIONS

Silicon Carbide Membranes for Water Filtration Applications 121
Melanie Kuhn, Abhaya Bakshi, Eric Sheridan, Fabiano Rodrigues, Adrien Vincent, Malte Moeller, and Ronald Neufert

Fabrication of Porous Ceramics with Cylindrical Pores and Incorporating Pores by Unidirectional Solidification Process 129
Shunkichi Ueno and Jun-Woo Lee

ADVANCED SENSORS FOR ENERGY, ENVIRONMENT, AND HEALTH APPLICATIONS

Printed Cantilevers and MOS Gas Sensors for Hazardous Gas Detection at Room Temperature 139
Hélène Debéda, Van Son Nguyen, Fernando Almazán, Maria Pina Pilar, Véronique Jubéra, and Claude Lucat

Sensing Characterization of the MOS Micro Gas Sensor Array on Gas Mixture 147
Bum-Joon Kim and Jung-Sik Kim

Author Index 159

Preface

The global challenges we face require innovative thinking and sustainable technology to meet increased demands for energy, clean water, and infrastructure. Research in materials, specifically ceramic materials, continues to provide solutions to everyday challenges such as environmental protection, energy supply and generation, and healthcare. The 11th International Symposium on Ceramic Materials and Components for Energy and Environmental Applications (11th CMCEE), held June 14–19, 2015 at the Hyatt Regency Vancouver in Vancouver, B.C., Canada, identified key challenges and opportunities for ceramic technologies to create sustainable materials.

This Ceramic Transactions volume contains papers submitted from the following five symposia held in Track 3: Ceramics for Environmental Systems:

- Photocatalysts for Energy and Environmental Applications
- Advanced Functional Materials, Devices, and Systems for the Environment
- Geopolymers, Inorganic Polymer Ceramics, and Sustainable Composites
- Macroporous Ceramics For Environmental and Energy Applications
- Advanced Sensors for Energy, Environment, and Health Applications

After a peer-review process, 14 papers were accepted for inclusion in this proceedings volume. The editors wish to extend their gratitude and appreciation to all the symposium co-organizers for their help and support, to all the authors for their cooperation and contributions, to all the participants and session chairs for their time and efforts, and to all the reviewers for their valuable comments and suggestions. We also acknowledge the organization and leadership provided by the meeting chairs, Mrityunjay Singh, Tatsuki Ohji, and Alexander Michaelis.

We hope these proceedings will serve as a useful resource for researchers and engineers working in the field of environmental science and technology.

LIANZHOU WANG, Shanghai Institute of Ceramics, China
NOBUHITO IMANAKA, Osaka University, Japan
WALTRAUD M. KRIVEN, University of Illinois at Urbana-Champaign, USA
MANABU FUKUSHIMA, AIST, Japan
GIRISH KALE, University of Leeds, UK

Photocatalysts for Energy and Environmental Applications

EFFECT OF STRUCTURAL PROPERTIES ON THE PHOTOELECTROCHEMICAL PERFORMANCE OF TiO_2 FILMS

A. K. Alves; A. C. Teloeken; F. A. Berutti; C. P. Bergmann
Postgraduate Program in Mining, Metallurgy and Materials (PPGE3M)
Universidade Federal do Rio Grande do Sul (UFRGS)
Porto Alegre, RS, Brazil

ABSTRACT

Semiconductors can be excited by exposure to radiation of a higher energy than the band gap and an energy-rich electron–hole pair is formed. This energy can be utilized electrically, to change the catalyst surface itself, or chemically. For photoelectrochemical (PEC) water-splitting, a light-sensitive semiconductor is commonly used as a photoelectrode. In PEC applications, titanium dioxide (TiO_2) is one of the most important and most widely used semiconductors, mainly due to its chemical stability, non-toxicity, low cost and adequate band gap for effective water-splitting. The production of TiO_2 thin films is a feasible way to immobilize the catalyst in a photoelectrode surface. The photoactivity of the resulting TiO_2 film depends strongly on its physical properties such as crystal phase structures, thickness, porosity and atomic defects. In the present work, these properties were studied for TiO_2 thin films obtained by a dip-coating process of a sol-gel system in a glass-FTO substrate. Two systems were tested; one with a binder (polyvinyl butyral – PVB) and another without it. The samples after heat treatment were characterized by XRD, SEM, DRS, ellipsometry, profilometry and photocurrent measurements. A significant correlation between the structural aspects of the films (roughness, thickness and optical properties) and the generated photocurrent was established.

INTRODUCTION

In the photoelectrochemical decomposition of water, which takes place in photoelectrochemical cells, hydrogen and oxygen are separately generated on the surface of the photocathode and the photoanode. Under illumination with light of a specific wavelength, holes (h^+) will be generated in the valence band, diffuse to the surface of the catalyst and then oxidize the water molecules of the medium producing O_2. The photogenerated electrons in the conduction band can be transferred to the photocathode via the electrolyte and generate H_2 [1]. The energy necessary for such a process to occur (thermodynamic potential) is limited by the electrochemical thermodynamics of the chemical decomposition of water into H_2 and O_2, typically requiring 1.23 V. However, when using just one semiconductor to absorb light, an electrochemical overpotential is necessary to overcome the kinetics barrier[1].

Titanium oxide has been used as a photoanode in this process because it has interesting energy band positions, chemical stability, non-toxicity and low cost[2]. Although TiO_2 has already been studied as a photoanode, the structural and surface characteristics of TiO_2 thin films that can effectively influence the efficiency of H_2 production are yet to be full comprehended. In general, the preparation conditions of TiO_2 thin films based on a sol-gel process can strongly affect the physical properties of the film[3, 4]. Therefore, it is necessary to systematically study the structural and physical properties of these films according to the preparation conditions.

In this context, this paper describes the preparation of TiO_2 thin films using a sol-gel dip-coating process. Titanium propoxide was used as a precursor and polyvinyl butyral (PVB) as a binder system to tailor the viscosity of the system. The aims of this work were to obtain TiO_2 films with a specific crystal structure (anatase) using a low processing temperature and, to study the influence of parameters such as the heat treatment temperature, the number of film layers, the aging time and the presence of PVB in the morphological, optical and photoelectrochemical characteristics of the films.

EXPERIMENTAL

A solution was prepared by mixing titanium propoxide (Sigma-Aldrich) and acetic acid (Sigma-Aldrich) (volume ratio 1:1) under magnetic stirring for 10 minutes. Following the approach adopted by Alves et al.[5], the sol was then kept for 15 minutes in the dark to complete the hydrolysis reaction. Subsequently 8 mL of anhydrous ethanol (Dinamica), 0.8 mL of acetylacetone (Sigma-Aldrich) and 0.1 mL of Triton X-100 (Sigma-Aldrich) were successively added.

To tailor the viscosity of the system, a second solution was prepared using the same procedure described above; however 0.316 g of polyvinyl butyral (Mowital B 30H, Omya) were dissolved in 8 mL of anhydrous ethanol and added to the other reagents already mixed.

The solutions were aged for 2, 4, 6 and 8 hours in the dark. Afterwards, each solution was deposited on fluorine-doped tin oxide (FTO) coated glass substrates (NSG TEC 8A, Xop Física) by a dip coating technique (Compact DipMaster 50 Dip Coater). The substrates were cleaned with acetone using an ultrasound bath for 10 minutes, rinsed with distilled water and ethanol, and dried in air for 30 minutes.

The substrates were dipped 1, 2 or 3 times into each solution at a speed of 50 mm/min, kept immersed for 30 seconds and removed with the same speed. After each coating step, the samples were dried for 30 minutes in air and, after the last coat, dried for 24 h at 100°C. The samples were then heat-treated at 400, 500 or 600°C, with a heating rate of 100 °C/h, and a dwelling time of 2 hours.

Characterization

Viscosity measurements (Rheolab MC 120, Anton Paar) of the prepared solutions with and without PVB were measured as a function of the shear rate (from 100 to 1000 s^{-1}) at different time intervals (2, 4, 6 and 8 h) to monitor the system aging and the effect of PVB.

The morphology of the films was analyzed using Scanning Electron Microscopy (SEM Hitachi TM3000). The samples were attached to a sample holder with a carbon tape, without any previous preparation.

The crystalline phase composition of the samples was determined using the X-ray diffraction technique (Philips X'Pert), Cu $K\alpha_1$ radiation (40 mA, 40 kV). The measurement was made in $\theta/2\theta$-configuration in a range of $5° \leq 2\theta \leq 80°$ with a scan speed of 2 s/step and an increment of 0.002°. The crystallite size of the samples was calculated using the Scherrer equation[6].

The optical properties of the films were determined using diffuse reflectance spectroscopy (DRS) with an integrating sphere (Agilent, Cary 5000) for band gap calculation and spectroscopic ellipsometry (SOPRA GES-5E IRSE, Xe light) or optical profilometry (Contour GT-K 3D, Bruker) for determining the film thickness of samples (opaque or translucid/transparent, respectively).

The roughness of the films was determined using a Rugosimeter SJ-400 (Mitutoyo), analyzing 0.01 μm^2 of the surface area of each photoelectrode. Three distinct areas of each sample were measured and the medium value of Ra was calculated.

Photocurrent measurements were performed using 1M KOH solution as electrolyte, a potentiostat (Autolab, Metrohm), a platinum counter electrode and an Ag/AgCl/3M-KCl reference electrode. A potential bias was applied from -0.5 to 0.5 V, at a scan rate of 10 mV/sec. The measurements were first performed in the dark and then under illumination using a solar simulator (Oriel Lamp by L.O.T – Oriel AG), at an intensity correspondent to 1.5 AM (1 sun, 1000 W/m^2).

RESULTS AND DISCUSSION

The aging affect on the viscosity of the systems is observed in the results shown in Figure 1. As a high molecular weight polymer, it was expected that the samples with PVB would present a higher viscosity than the samples without it, where the vehicle is mainly ethanol. The aging effect is more distinctive in the first two hours, and is possibly caused by the condensation reactions of the sol-gel system[7]. Both systems have an almost Newtonian fluid behavior.

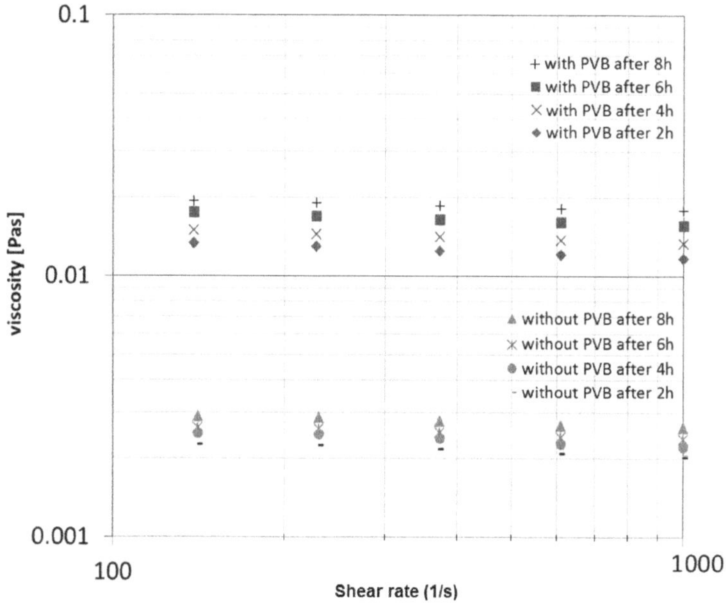

Figure 1. Effect of aging time in the viscosity of the produce solutions with and without PVB.

It has been reported[8] that it is possible to directly correlate the thickness of a dip-coating film to the viscosity of the system. In fact, when analyzing the thickness results (Table 1) it is possible to observe that the systems that contain PVB, the more viscous solutions, produce thicker films. For the systems without PVB, which have a very low viscosity, the film thickness was on average 0.14 nm. Sonawane et al.[9] reported that when depositing multiple times, solutions with low viscosity have adherence difficulties with respect to the last layer and, non-uniform films were obtained. This may explain why the films obtained without PVB showed practically no change in thickness with an increase in the number of layers. On the other hand, as also observed by Sonawane et al.[9], the addition of a polymer system helps to improve the adhesion of a new layer and thus increase the thickness with the deposition of multiple layers.

Table 1. Thickness, roughness and band gap of the produced films heat-treated at 500°C.

Sample	Thickness [nm]	Roughness (Ra) [nm]	Band gap [eV]
1 layer	0.12	0.25	2.20
2 layers	0.13	0.30	2.66
3 layers	0.16	0.53	2.71
1 layer with PVB	0.23	0.35	2.97
2 layers with PVB	0.33	0.45	3.42
3 layers with PVB	0.42	0.82	3.24

Nevertheless, the number of layers affects the roughness of films. Films with thicker layers, especially those with PVB, have surfaces that are more irregular, probably due to the thermal decomposition of the organic binder. When this exits the system in the gaseous phase, it leaves cracks and mesopores in the film. The surface images obtained by SEM for films without and with PVB, (Figures 2 and 3, respectively), corroborate the roughness measurements.

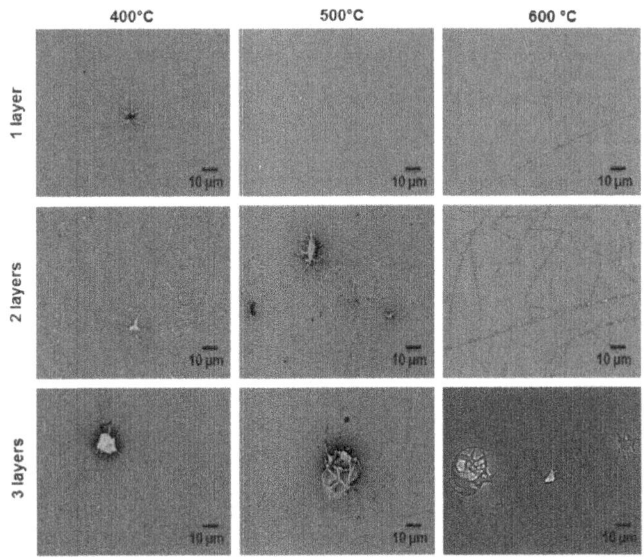

Figure 2. SEM images of films with 1, 2 or 3 layers, without PVB.

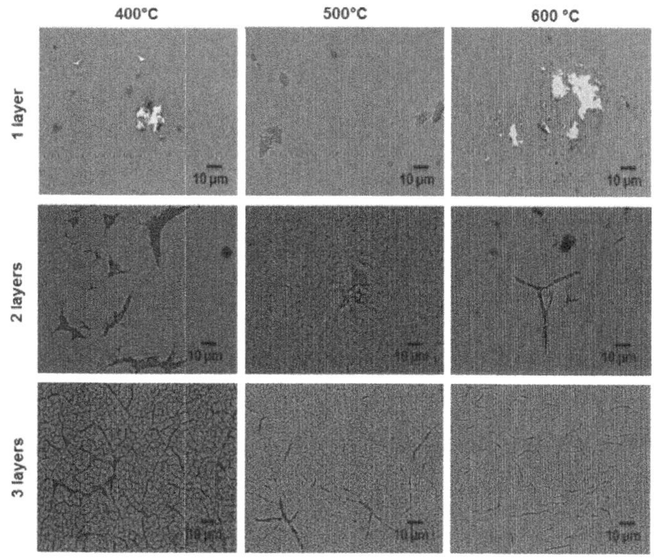

Figure 3. SEM images of films with 1, 2 or 3 layers, with PVB.

Films obtained without the binder systems present a more uniform and homogeneous surface, although some cracks are identified, especially at higher treatment temperatures, and are probably caused by differences in the thermal expansion of the substrate and the film. However, in the case of films that contain PVB, the microstructure is completely different. Many large and small cracks are clearly visible, particularly for samples with three layers. In these systems, the films are white/translucent, an indication of the presence of many structural defects (microporosities), greater thickness and higher amounts of TiO_2 particles in the surface.

The heat treatment temperature has no significant effect on the thickness, surface roughness or phase composition. The XRD results presented in Figure 4 show that only the anatase phase was identified as the major component of the films with PVB heat-treated at 400, 500 or 600°C. The crystalline size increases with the treatment temperature, but the presence of PVB does not significantly influence this property, since its decomposition occurs before the phase crystallization start[10, 11].

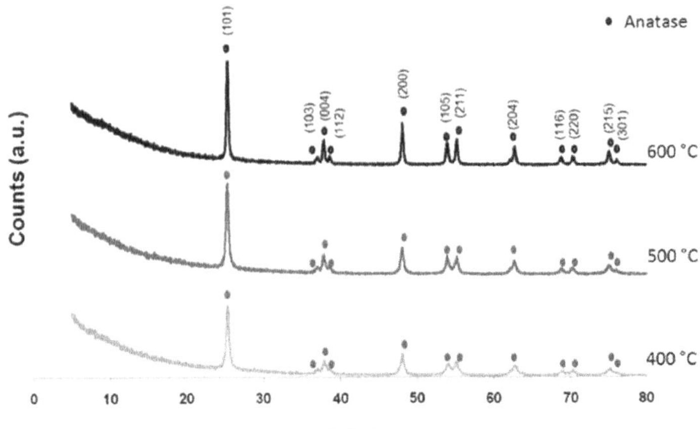

Figure 4. XRD analysis of the samples with PVB heat-treated at 400, 500, 600°C.

The crystallite size of the particles in the films (Table 2) was not affected by the number of deposited layers, but higher heat treatment temperatures led to grain growth. The effect of crystallite size on the photocurrent of the films can be observed in Figure 5, where samples with three layers of the films with PVB were tested. The photocurrent increases with an increase in the crystallite sizes. Usually, to achieve the best performance in terms of photocurrent generated by solar irradiation, the crystallite sizes and film thicknesses are important factors[12]. To maximize electron/hole separation efficiency, the crystallite sizes should be smaller than the diffusion length of the carriers but, on the other hand, as crystallites become smaller, more interfaces are created, which decrease the carriers' mobility and ultimately limits photocurrent generation[12].

Table 2. Crystallite size of the particles obtained from systems with and without PVB as a function of the heat-treatment temperature.

Heat treatment temperature (°C)	Crystallite size (nm)	
	Without PVB	With PVB
400	19.4	17.9
500	28.4	26.3
600	43.0	45.8

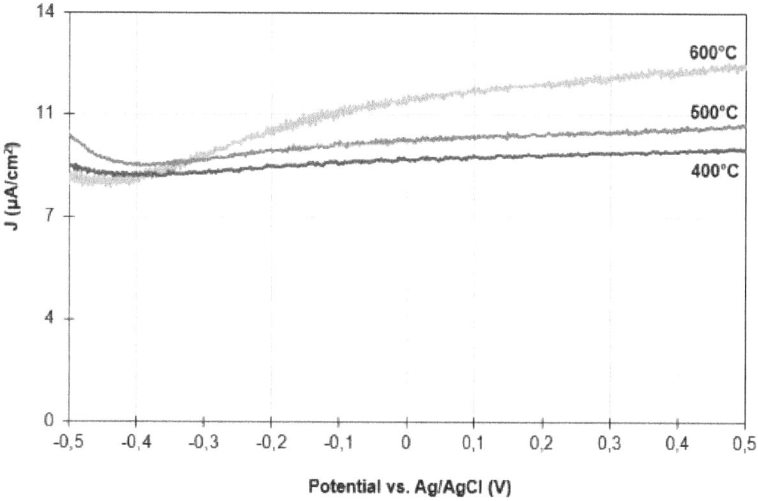

Figure 5. Photocurrent of the 3 layer films with PVB heat treated at 400, 500 or 600°C.

Additionally, thicker layers absorb more incident photons[13]. A correlation was observed between the photocurrent (Figure 6) and the microstructural properties of the films. The number of layers had a significant effect on samples that contain the binder, but no direct effect on samples without it. Since the thicknesses of the films without PVB are very similar, the amount of particles in the surface are practically the same, and thus the photocurrent values are of the same order of magnitude. Films with two and three layers appear to have smaller photocurrents. This effect may be explained by adherence and dissolution difficulties of these layers when formed from very low viscosity systems[9].

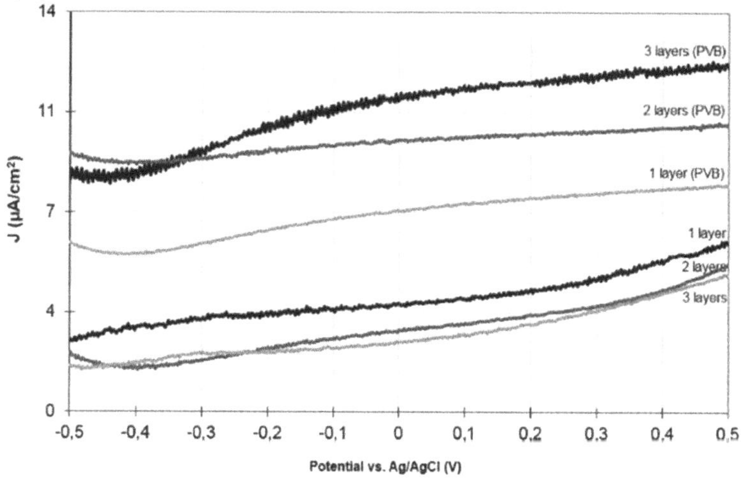

Figure 6. Photocurrent measurements of samples with and without PVB, with 1, 2 or 3 layers.

It can also be observed that the photocurrent values were improved significantly with the addition of PVB to the system. When examined by SEM, samples with PVB have an important microstructural aspect: mesoporosity, which is crucial to ensure penetration of the electrolyte within the photoanode and thus aiding electron transfer. According to optical band gap analysis, the higher photocurrent observed could be attributed to higher band gaps shown by these samples. In addition, the binder had a significant influence on the coating thickness, which is generally one of the main parameters for increasing the photocurrent in a TiO_2-based photoelectrodes, since more photoactive material is available to be sensitized by the incident light.

CONCLUSIONS

The influence of the microstructural, optical and photoelectrochemical characteristics of multilayer TiO_2 thin films obtained using a dip-coating technique, with and without a binder system, was investigated. The presence of PVB as a binder promotes an increase in the thickness and roughness of the films, as well as opacity and photocurrent.

Using the proposed synthesis method, the anatase phase was obtained for films with or without PVB up to 600°C. The crystallite sized increased with the temperature of heat treatment, but there was no effect of the binder on this property. The photocurrent is also sensitive to crystallite size and increases with their growth.

The aging time slightly increases the viscosity of the systems, following an apparent Newtonian behavior pattern. However, the presence of PVB significantly increases their viscosity and thus, when dip-coated, the produced layers are thicker and rougher. This leads to the formation of a mesoporous structure that is of the upmost importance in PEC, due to the penetration of the electrolyte into the electrode and effectively facilitates electron transfer, as observed by the higher photocurrents in films obtained using the binder.

ACKNOWLEDGEMENTS

The authors would like to thank PRH-ANP, CNPQ, CAPES and UFRGS for their support.

REFERENCES

[1] Pu, Y-C., Ling, Y., Chang, K-D., Liu, C-m., Zhang, J. Z., Hsu, Y-J., & Li, Y. (2014) Surface Passivation of TiO_2 Nanowires Using a Facile Precursor Treatment Approach for Photoelectrochemical Water Oxidation. *The Journal of Physical Chemistry,* Volume 118, 15086-15094.

[2] Li, Y., Yu, H., Zhang, C., Fu, L., Li, G., Shao, Z., & Yi, B. (2013) Enhancement of photoelectrochemical response by Au modified in TiO_2 nanorods. *International Journal of Hydrogen Energy,* Volume 38, 13023-13030.

[3] Lewkowicz, A., Synak, A., Grobeln, B., Bojarski, P., Bogdanowicz, R., Karczewski, J., Szczodrowski, K., & Behrendt, M. (2014) Thickness and structure change of titanium (IV) oxide thin films synthesized by the sol–gel spin coating method. *Optical Materials,* Volume 36, 1739-1744.

[4] Borlaf, M., Colomer, M. T., Moreno, R., & Ortiz, A. L. (2014) Rare earth-doped TiO_2 nanocrystalline thin films: Preparation and thermal stability. *Journal of the European Ceramic Society,* Volume 34, 4457-4462.

[5] Alves, A. K., Berutti, F.A, Clemens, F. J., Graule, T., & Bergmann, C. P. (2009) Photocatalytic activity of titania fibers obtained by electrospinning. Materials Research Bulletin, Volume 44, 312-317.

[6] Patterson, A.L. (1939) The Scherrer Formula for X-Ray Particle Size Determination. *Physical Review*, 56, 978-982.
[7] Chen K. F., Liu, C-H., Hsieh, C-K., Lin, C-L., Huang H-K., Tsai, C-H., & Chen, F-R. (2014) New fabrication process of long-life dye-sensitized solar cells by in situ gelation of quasi-solid polymer electrolytes. *Journal of Power Sources*, Volume 247, 939-946.
[8] Cisneros-Zevallos, L., & Krochta, J.M. (2003) Dependence of Coating Thickness on Viscosity of Coating Solution Applied to Fruits and Vegetables by Dipping Method. *Journal of Food Science*, Volume 68, 503-510.
[9] Sonawane, R. S., Kale, B. B., & Dongare, M. K. (2004) Preparation and photo-catalytic activity of Fe-TiO_2 thin films prepared by sol–gel dip coating. *Materials Chemistry and Physics*, Volume 85, 52-57.
[10] Dhaliwal, A. K., & Hay, J. N. (2002) The characterization of polyvinyl butyral by thermal analysis. *Thermochimica Acta*, Volume 391, 245-255.
[11] Courtecuisse, V. G., Chhor, K., Bocquet, J-F., & Pommier, C. (1996) Kinetics of the Titanium Isopropoxide Decomposition in Supercritical Isopropyl Alcohol. *Industrial & Engineering Chemistry*, Volume 35, 2539-2545.
[12] Sinclair, T. S., Hunter, B. M., Winkler, J. R., Gray, H. B., & Müller, A. M. (2015) Factors affecting bismuth vanadate photoelectrochemical performance. *Materials Horizons*, Volume 2, 330-337.

PHOTOCATALYTIC DEGRADATION OF DYES USING MWCNT-TiO_2 COMPOSITES AS CATALYST

F. A. Berutti; A. P. Garcia; A. K. Alves; S. Da Dalt; C. P. Bergmann
Postgraduate Program in Mining, Metallurgy and Materials (PPGE3M)
Universidade Federal do Rio Grande do Sul (UFRGS)
Porto Alegre, RS, Brazil

ABSTRACT

Synthetic dyes are the major pollutants in textile industries wastewaters. Because of their toxicity and slow degradation, these dyes are environmentally hazardous materials. An efficient method for the decomposition of dyes is photocatalytic process. In order to optimize the catalytic properties of semiconductors applied in photocatalysis, MWCNTs have been used as support to help the electron transfer process and, therefore, increasing the photocatalytic activity. In this context, multiwall carbon nanotubes-TiO_2 composites (MWCNT-TiO_2) were prepared using titanium propoxide and commercial TiO_2 as titanium oxide sources, to degrade a methyl orange dye through photocatalysis using UV irradiation (λ=365 nm). The composites were prepared by solution processing followed by a thermal treatment at 400, 500 or 600°C. The material was characterized using XRD, N_2 adsorption (BET method) for specific surface area determination, TEM for microstructural analysis and the optical characteristics were evaluated using photoluminescence and diffuse reflectance techniques. It was observed that the CNT have two distinct activities, one is the adsorption of the dye in its surface and the second the formation of a heterojunction with TiO_2 to promote higher photoactivities. A correlation between the structural atomic defects and the photoactivity of the MWCNTs-TiO_2 composites was found. The excess of MWCNT and or TiO_2 particles in the reaction medium promotes the scattering of the light and lower photodecomposition rates.

INTRODUCTION

Composite materials consisting of CNTs and semiconductors with photocatalytic properties have attracted the attention of researchers due to the application of these materials for water and air contamination treatment, hydrogen evolution, photoreduction of CO_2, dye sensitized solar cells and other applications.

In order to improve the catalytic properties of semiconductor oxides such as TiO_2, SnO_2, ZnO, WO_3, to name a few, CNTs have been used as support, which can facilitate the electron transfer, while slowing down the electron-hole pair recombination, favoring the catalytic activity of these composites[1, 2].

However, some problems still hinder a greater efficiency of this photocatalytic semiconductor oxides attached to CNTs. Among them, it is possible to point out two: the intensity weakening of light that reaches the surface of the catalyst due to the high light absorption of the CNTs and, the lack of reproducibility during synthesis of the composite and further treatments[3]. In spite of that, the electric properties of CNTs are sensitive to surface load transfer and electrostatic changes in the surrounding environment, suffering drastic change by simple adsorption certain molecules[4], thus the pH of the medium can influence its photoactivity.

Among the photocatalyst semiconductor metal oxides, titanium dioxide present itself as attractive choice because of its high oxidizing property, absence of toxicity and long term photostability. As an n-type semiconductor, its conductivity increases with the degree of oxygen loss within the crystal lattice. Point defects such as oxygen vacancies (double charged) and interstitial titanium ions (triple charged) are the most common defects found in TiO_2 lattice[5].

Usually, oxygen vacancies are common under weakly reducing conditions or low annealing temperatures (below 600°C in vacuum), while in more reducing environments and higher annealing temperatures (above 800°C in vacuum), titanium interstitials become the predominant deffect[6,7]. It is also believed that the dispersion of TiO_2 on the surface MWCNTs favors the development of many active sites for photocatalytic degradation.

In this context, this work aims to correlate the photocatalytic performance of TiO_2 doped MWCNTs nanocomposite obtained using a solution route with titanium propoxide and a commercial powder mixture (P25 Evonik) annealed at different temperatures and under different pH conditions to its structural and morphological characteristics. Absorption characteristics of the composites in the dark was also evaluated.

EXPERIMENTAL

The functionalization of MWCNT (Baytubes, Bayer) was made following a sol-gel modified route. Firstly, 0,6 mol% of Titanium from Titanium (IV) propoxide (Sigma-Aldrich) was hydrolyzed with 0,1 mol% of nitric acid (Synth), following the addition of a mixture of 14 mol% of isopropanol (Dinamica) and 0,5 mol% of deionized water (140:5) under magnetic stirring for 1 h at 40°C in the dark. Additionally, 0.33 g of MWCNT was mixed with 29 mL of the titanium solution and stirred for another hour. This process yield an acidic reaction medium (pH 1.0). For synthesis in alkaline medium (pH 10) 0,2 mol% of ammonium hydroxide (Vetec) was added dropwise in the final step.

Similar procedure was employed using commercially available TiO_2 (AEROXIDE – P25, Evonik) replacing the titanium propoxide. This final mixture was kept under magnetic stirring before the system transforms into a gel. The resulting suspension was vacuum filtered and the solid material was dried at 100°C for 24h.

The material obtained using titanium propoxide was annealed at 400, 500 or 600°C for one hour at a heating rate of 2.5 °C/min, to promote phase crystallization. These samples were subdivided in two groups: the acid group (TA-400, TA-500 and TA-600) and the alkaline group (TB-400, TB-500 and TB-600). Samples obtained using P25, in acidic conditions (P25+CNT-A) and alkaline conditions (P25+CNT-B), were not subjected to annealing.

The phase composition was evaluated using the X-ray diffraction (XRD) technique (PanAnalytical, X'Pert MPD) equipped with a graphite monochromator and a copper anode, operating at 40 kV and 40 mA, in a 2θ range of 20-70°, steps of 0.05° for 2 s, with Cu $K\alpha$ radiation. The morphology was analyzed using scanning electron microscopy (SEM – JEOL JSM-6060) operating at 15 kV, and transmission electronic microscopy (TEM – JEM 1200EXII-120 kV) operating to 80 kV. The Brunauer–Emmett–Teller (BET) method, using N_2 as gas adsorption media (Quantachrome Autosorb NOVA 1000e) was used to determine the specific surface area. Optical characteristics (band gap) were determined using diffuse reflectance (Cary 5000 with integrating sphere, DRA-1800) and photoluminescence measurements (Princeton Instruments SP-2300i) were made under 266 nm UV laser irradiation of the samples, detected by a CCD cam (Pixis 256BR).

The photocatalytic activity of the obtained samples were evaluated following the decrease in the concentration of an aqueous solution containing 20 ppm of methyl orange, under UV light irradiation (λ = 365 nm; 3,397 eV). Initially, a sample (0.8 g/L) was mixture with 125 mL of the dye solution using an ultrasound probe for 10 min and kept it in the dark for 1h. Subsequently, the solution was transferred to a glass reactor where the reaction temperature was controlled by a thermostatic bath (30°C) and irradiated with UV light (UV dose: 60 mJ/cm^2 in the reactor interior). Samples were collected periodically from the reactor every 10 min during the time interval of 70 min and filtered to a 4 mL PMMA cuvette. The absorbance at the wavelength of 465 nm was recorded (Agilent, Cary 7000) for each sample.

RESULTS AND DISCUSSION

Figure 1 a and b presents the XRD diffraction patterns of the samples synthesized with P25 in acidic and alkaline medium and, titanium propoxide in acid conditions heat treated at different temperatures, respectively. Samples of pure P25 and pure MWCNT were also subjected to XRD analysis (Figure 1 a). The results show that P25 is composed of anatase and rutile, as reported previously[8]. When it was mixture with MWCNT there was an apparent decrease in its crystallinity, especially for the sample synthesized at acidic conditions (P25+MWCNT-A). Similar observation is possible when analyzing the crystallinity behavior of MWCNT. According to Wang et al.[9] the decrease in the XRD peak intensity of MWCNT is common when TiO_2 particles cover homogenously the MWCNT.

On the other hand, when titanium propoxide was mixture with MWCNT and then heat-treated (Figure 1 b), it was possible to identify anatase with low crystallinity as the major constituent of samples, independently of temperature and synthesis medium (acid or alkaline). The apparent low intensity of the anatase peaks can be associated with the presence of CNT that favors lower crystallization[9]. The absence of peaks indicating the presence of MWCNT at 25.2° might be attributed to the overlapping with the main peak of the anatase phase at 25.3°. The CNT peak at 44° weakens with the increasing of the heat treatment temperature, disappearing at 600°C (TA-600) due to decompositions process of the carbonaceous material.

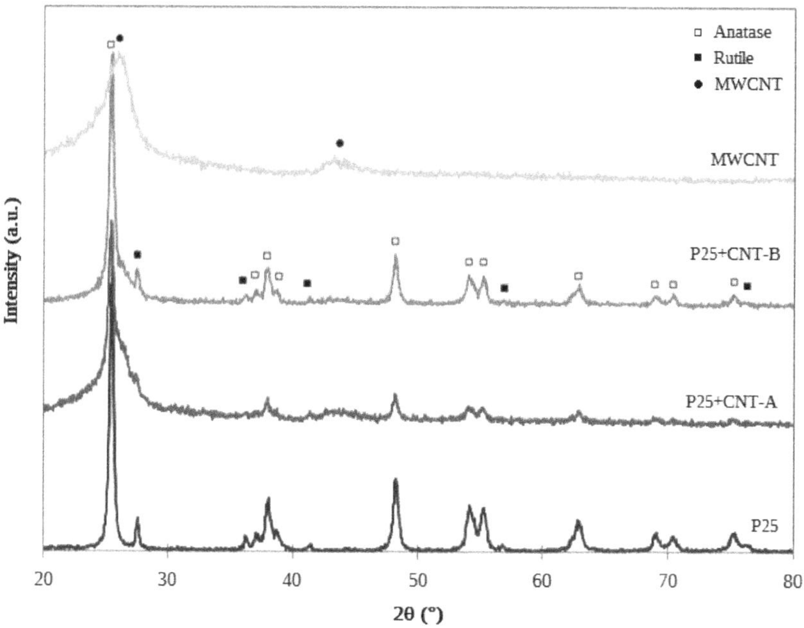

(a)

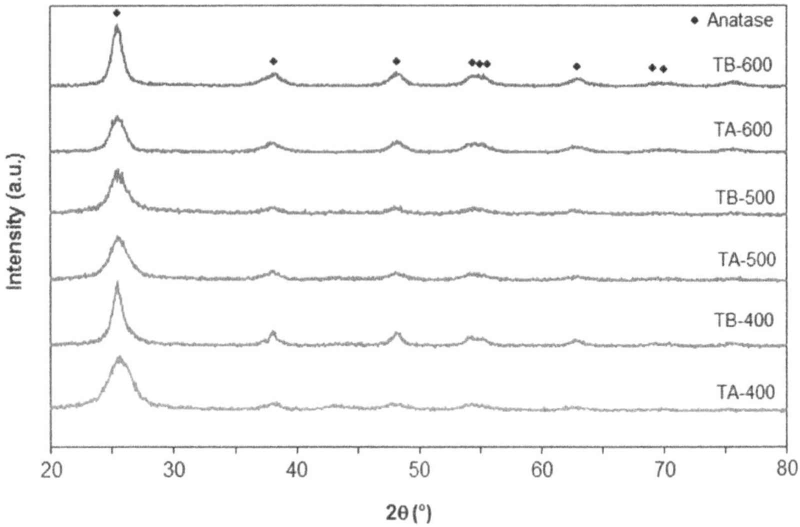

(b)

Figure 1. XRD diffraction patterns of the samples synthesized with P25 (a) and titanium propoxide in acidic (TA) or alkaline (TB) medium heat-treated at different temperatures (b).

Morphological observation by TEM (Figure 2 a, b and c) revealed the typical morphology and particle distribution of the samples prepared with titanium propoxide (TA-500 and TB-400) and with P25 (P25+CNT-A). Apparently, the TiO_2 particles in the samples TA-500 and TB-400 are spherical with diameters 30 nm and 10 nm, respectively. P25 particles (av. diam. 50 nm) are expected to be physically attached to the MWCNT due to the absence of heat treatment. The acidic medium used to obtain the TA-500 and P25+CNT-A samples seems to contribute to a uniform distribution of TiO_2 particles when compared to the agglomeration observed on the sample obtained in alkaline medium (TB-400).

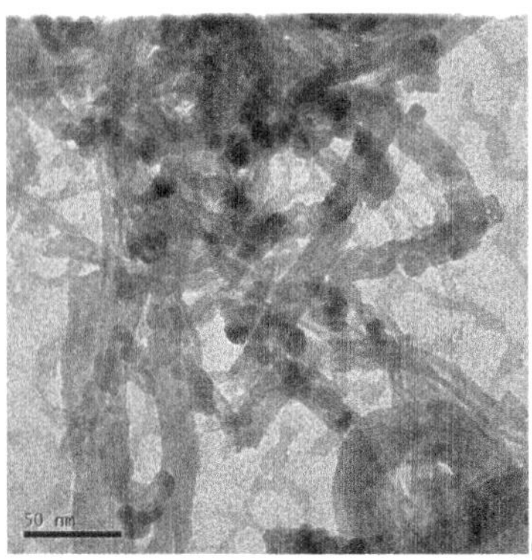

(a)

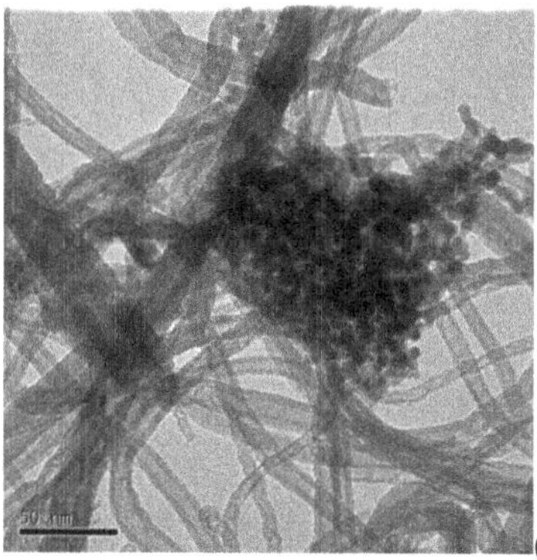
(b)

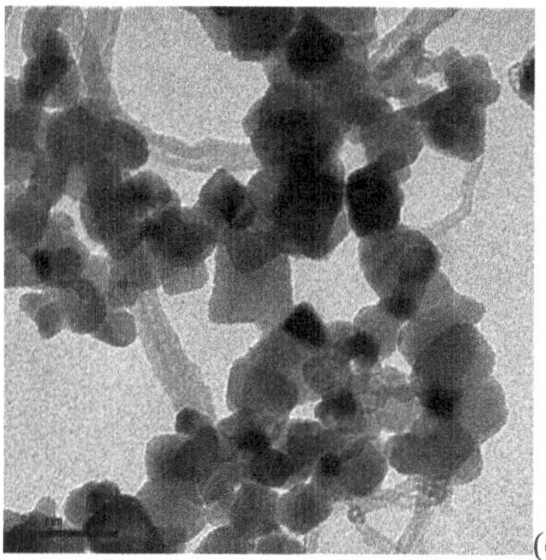
(c)

Figure 2. TEM images of samples (a) TA-500, (b) TB-400 and (c) P25+CNT-A.

Table 1 presents the values of band gap and specific surface area of the synthesized samples. Samples P25+CNT-A, TA-500, TB-400, TB-500 present a redshift if compared to P25, i.e., the band gap values of these samples are lower than the band gap of P25. This is an indicative that the presence of MWCNT reduces the necessary energy to electron transition between bands, contributing to the quantum efficiency during photocatalysis [10, 11].

For samples synthesized using titanium propoxide in acidic conditions and heat-treated at 400, 500 and 600°C it is possible to observe an increase in the band gap value with the increase of the treatment temperature. It is caused probably, by the development of the structural organization of the crystalline network due to the heat treatment process (phase crystallization)[10].

Regarding the specific surface area of the samples synthesized with titanium propoxide in acid or alkaline environment, there is an increase in its value when the temperature reaches 500°C. This is probably due to a more intense decomposition of organic materials at higher treatment temperatures. Some authors[12,13] suggest that the presence of high amounts of MWCNT in the composites induces an increase in the specific surface area and a higher particle dispersion.

A reduction of the surface area at higher temperatures (600°C) can be associated to two effects: (i) sintering of TiO_2 particles, as a result of the natural process to reach the lowest energy level of the particulate system and, (ii) the reduction of the amount of MWCNT in the sample due to thermal decomposition at higher temperature.

Table 1. Band gap and specific surface area of MWCNT, P25 and TiO_2 composites.

Sample	Band gap (eV)	Specific surface area (m^2/g)
MWCNT	---	225,5
P25	3,53	68,1
P25+CNT-A	3,42	63,8
P25+CNT-B	3,58	143,3
TA-400	3,07	122,3
TA-500	3,20	194,0
TA-600	3,55	150,9
TB-400	3,52	181,2
TB-500	3,47	211,7
TB-600	3,64	158,5

The photoluminescence (PL) analysis of MWCNT, P25 and their composites (P25+CNT-A and P25+CNT-B) and the photocatalytic activity of these samples are presented in Figures 3 and 4, respectively.

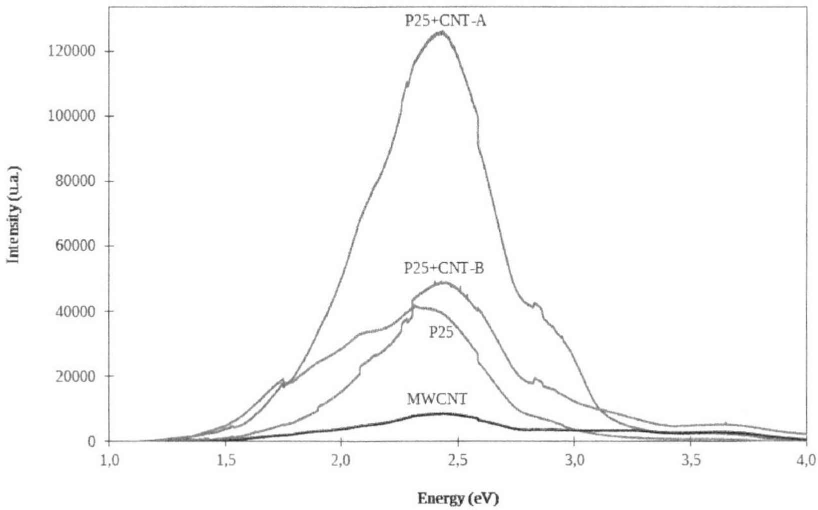

Figure 3. Photoluminescence of MWCNT, P25 and composites P25+CNT-A and P25+CNT-B.

The PL of the P25 sample has a broad emission band, with a peak around 520 nm (2.38 eV) associated with anatase, similar to the results of Fujihata et al.[14]. There is a redshift of the PL spectrum of this sample when compared to its CNT composites. It suggests that the P25 PL emission occurs at a lower energy level, which can be attributed to a lower density of defects[15]. Due to the TiO_2 agglomeration observed in basic medium, the PL of sample P25+CNT-B has a lower intensity than the sample obtained in acid environment (P25+CNT-A). Acording to the literature[16], agglomeration increases the surface recombination traps, decrease the quantum yield and the decrease in the value of the PL peak.When the MWCNT are homogenously covered with TiO_2 particles, it is expected that the PL signal of the composite will be higher that the MWCNT alone, due to the presence of TiO_2 particles with stronger PL signal.

The photocatalytic activity of these samples indicates that in the case of pure MWCNT, only adsorption effects are present. The high surface area of the CNT promotes the presence of actives sites for the complete adsorption of the dye before the photocatalytic test began. In the case of the sample P25+CNT-B, due to the agglomeration cause by the basicity of the reaction environment, larger amounts of CNT surface area was uncovered by TiO_2 particles. Because of that, there was a complete absorption of the dye, with a 10-minute delay when compared with pristine CNT, due the presence of TiO_2 particles, which lower its available surface area. Probably because the absorption effects due to the presence of CNT, the composites seams to present a better photocatalytic activity.

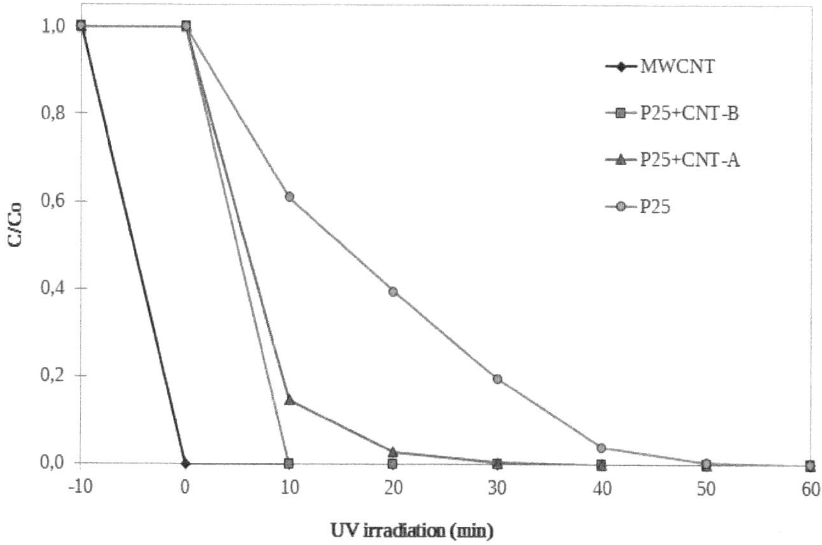

Figure 4. Variation of the concentration of the dye under 60 min period of UV irradiation in the presence of MWCNT, P25 and composites P25+CNT-A and P25+CNT-B.

On the other hand, the PL of the composite samples prepared through the heat treatment of titanium propoxide in acid and alkaline medium (TA and TB) (Figure 5) presents emission bands at 325 and 380 nm (3,81 and 3,26), i.e., more energetic vibrational modes. The anatase PL peak (2.38 eV) is broader and presents a blueshift, an indicative of high oxygen-vacancy density[17, 18]. The blueshift is associated with TiO_2 agglomeration[19], specially observed by TEM for alkaline conditions, which reduces the reactive surface area available. Both assumptions might explain the low photoactivity of TB-400 and TB-600 composites.

The PL spectra revealed a broad asymmetric band from UV (4.0 eV) to near-infrared (1.2 eV). The UV emissions are originated from interband electrons transitions[20]. In the visible region the emission is due to the recombination of the electron in the oxygen vacancy with a hole[21]. This is directly associated with the photoactivity (Figure 6), when this recombination should be avoided. The lower photoactivity of the composites TA-600 and TB-400, TB-500 and TB-600 may be associated to this effect. The photoactivity of the composites TA-400 and TA-500 have a strong decrease probably caused by adsorption phenomena of the dye on the surface of the CNT that overcome the photocatalytic effect.

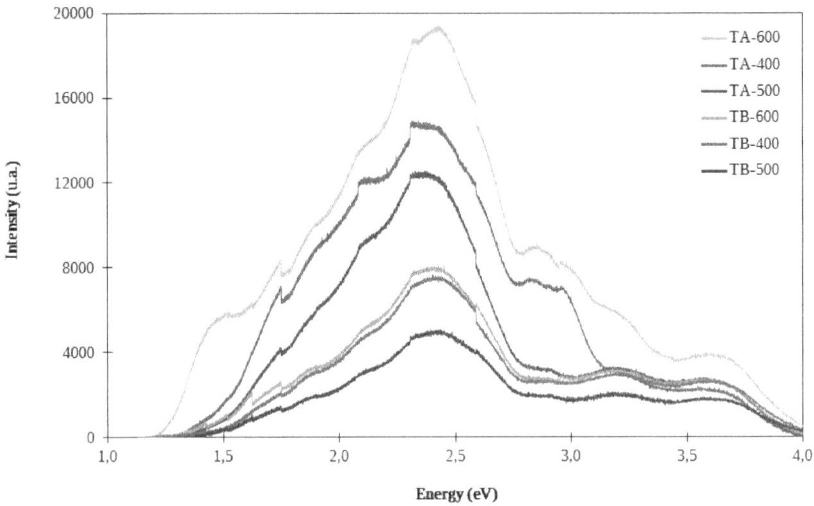

Figure 5. Photoluminescence of TiO_2 composites heat-treated at different temperatures in acid and alkaline medium.

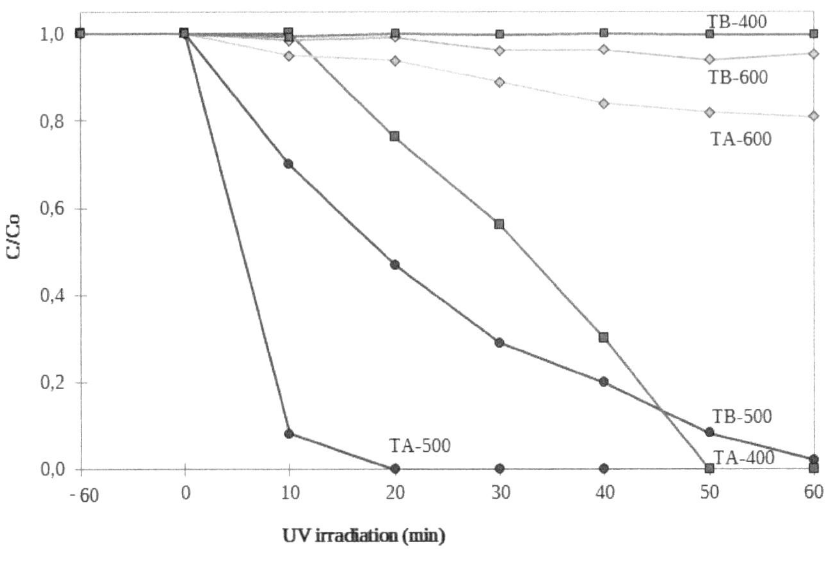

Figure 6. Variation of the concentration of the dye under 60 min period of UV irradiation in the presence of TiO_2 composites heat-treated at different temperatures in acid and alkaline medium.

CONCLUSIONS

It was possible to obtained TiO_2-MWCNT composites using a solution technique using P25 and titanium propoxide (followed by thermal treatment) under acid or alkaline medium. The alkaline environment promotes agglomeration of TiO_2 particles hinder the photocatalytic activity.

The presence of MWCNT in the composites favors an increase of the specific surface area and consequently the absorption phenomena. The presence of CNT can also facilitate the charge transfer in the semiconductor and difficult the electron/hole recombination.

The band gap of the composites that are higher than the energy of the UV light used is associated with a very low photoactivity. In those cases, the UV energy was unable to promote the charge transfer within the composite. A higher photoactivity was associated with composites with smaller band gaps.

The role of the electronic defects is also fundamental to understand the photocatalytic behavior of the TiO_2-CNT composites. In fact, the increase of the PL emission is associated to defects in the TiO_2 structure, originating intermediate energy states or oxygen vacancies. Higher heat treatment temperatures in air atmosphere cause an increase of the oxygen defects, originating less photoactive composites.

ACKNOWLEDGEMENTS

The authors would like to thank PRH-ANP, CNPQ, CAPES and UFRGS for their support.

REFERENCES

[1] Leary R., & Westwood A. (2011). Carbonaceous nanomaterials for the enhancement of TiO_2 Photocatalysis. *Carbon*, Volume 49, 741-772.

[2] Majumdar, S., Nag, P., & Devi, P. S. (2014). Enhanced performance of CNT/SnO_2 thick film gas sensors towards hydrogen. *Materials Chemistry and Physics*, Volume 147, 79-85.

[3] Park, C. Y., Kefayat, U., Vikram, N., Ghosh, T., Oh, W. C., & Cho, K. Y. (2013) Preparation of novel CdS-graphene/TiO_2 composites with high photocatalytic activity for methylene blue dye under visible light. *Bulletin of Materials Science*, Volume 36, 869-876.

[4] Shim, M., Kam, N. W. S., Chen, R. J., Li, Y., & Dai, H. (2002). Functionalization of Carbon Nanotubes for Biocompatibility and Biomolecular Recognition. *Nano Letters*, Volume. 2, 285-288.

[5] Arghya Narayan Banerjee (2011) The design, fabrication, and photocatalytic utility of nanostructured semiconductors: focus on TiO_2-based nanostructures. *Nanotechnology, Science and Applications*, Volume 4, 35-65.

[6] Ardakani, H. K.(1994) Electrical and optical properties of in situ "hydrogenreduced" titanium dioxide thin films deposited by pulsed excimer laser ablation. *Thin Solid Films*, Volume 248, 234-239.

[7] Kofstad P. (1983) Nonstoichiometry, Diffusion, and Electrical Conductivity in Binary Metal Oxides. John Wiley and Sons, Inc., New York, NY.

[8] Rui, Z., Wu, S., Peng, C., & Ji, H. (2014) Comparison of TiO_2 Degussa P25 with anatase and rutile crystalline phases for methane combustion. *Chemical Engineering Journal*, Volume 243, 254-264.

[9] Wang, W., Serp, P., Kalck, P., & Faria, J. L. (2008) Preparation and characterization of nanostructured MWCNT-TiO_2 composite materials for photocatalytic water treatment applications. *Materials Research Bulletin*, Volume 43, 958-967.
[10] Zhang J., Huang Z-H, Xu Y., Kang F-Y (2011) Carbon-coated TiO_2 composites for the photocatalytic degradation of low concentration benzene. *New Carbon Materials*, Volume 26, 63-70.
[11] Fan, H. J. (2011) Mechanistic pathways differences between P25-TiO_2 and Pt-TiO_2 mediated CV photodegradation. *Journal of Hazardous Materials*, Volume 185, 227-235.
[12] Wang F., Zhang K. (2011) Reduced grapheme oxide-TiO_2 nanocomposite with high photocatalystic activity for the degradation of rhodamine B. *Journal of Molecular Catalysis A: Chemical*, Volume 345, 101-107.
[13] Sampaio, M. J., Silva, C. G., Marques, R. R. N., Silva, A. M. T., Faria, J. L. (2011) Carbon nanotube-TiO_2 thin films for photocatalytic applications. *Catalysis Today*, Volume 161, 91- 96.
[14] Fujihara, K., Izumi, S., Ohno, T., & Matsumura, M. (2000) Time-resolved photoluminescence of particulate TiO2 photocatalysts suspended in aqueous solutions. *Journal of Photochemistry and Photobiology A: Chemistry*, Volume 132, 99-104.
[15] Xin, Y., Lin, T., Li, S., Ling, Z., Liu, G., Huang, Z., & Lin, J. (2012) Preparation and photoluminescence of single conjugated polymer–TiO_2 composite nanofibers. *Journal of Luminescence*, Volume 132, 738-742.
[16] Li, Y., Giles, A., & Stokes, B. (2010) Photoluminescence of CdSe/ZnS quantum dots in GaN/quantum dots/GaN wafer bonded system. *Integrated Optoelectronics 5*, Volume 28, 19-25.
[17] Barberio, M., Barone, P., Pingitore, V., & Bonanno, A. (2012) Optical properties of TiO_2 anatase - Carbon nanotubes composites studied by cathodoluminescence spectroscopy. *Superlattices and Microstructures*, Volume 51, 177-183.
[18] Vishwas, M., Rao, N. K., & Chakradhar, R.P.S. (2012) Influence of annealing temperature on Raman and photoluminescence spectra of electron beam evaporated TiO_2 thin films. *Spectrochimica Acta Part A: Molecular and Biomolecular Spectroscopy*, Volume 99, 33-36.
[19] Xu, L., Shen, H., Li, X., & Zhu, R. (2010) Influence of annealing temperature on the photoluminescence property of ZnO thin film covered by TiO2 nanoparticles. *Journal of Luminescence*, Volume 130, 2123-2127.
[20] Srivastava, A. K., Deepa, M., Bhandari, S., & Fuess, H. (2008) Tunable Nanostructures and Crystal Structures in Titanium Oxide Films. *Nanoscale Research Letters*, Volume 4, 54-62.
[21] Silva, M. S., Cilense, M., Orhan, E., Góes, M. S., Machado, M.A.C., Santos, L. P. S., Paiva-Santos, C. O., Longo, E., Varela, J.A., Zaghete, M.A., & Pizani, P. S. (2005) The nature of the photoluminescence in amorphized PZT. *Journal of Luminescence*, Volume 111, 205-213.

SYNTHESIS OF THE TiO_2-LONG LASTING PHOSPHOR ($Sr_4Al_{14}O_{25}$:Eu^{2+},Dy^{3+}) COMPOSITE AND ITS PHOTOCATALYTIC REACTION PROPERTIES

Jung-Sik Kim[1*], Hyun-Je Sung[1], and Sang-Chul Jung[2]
[1]Department of Materials Science and Engineering, The University of Seoul, Seoul, 130-743 Korea
[2]Department of Environmental Engineering, Sunchon National University, Jeonnam 540-742 Korea

ABSTRACT

This study has investigated on the photocatalytic behavior of the nanocrystalline TiO_2 coupled with long lasting phosphorescent materials. Nanocrystalline titanium dioxide layers were directly deposited on the long lasting phosphor plates of strontium aluminates, $Sr_4Al_{14}O_{25}$:Eu^{2+},Dy^{3+}, by low-pressure chemical vapor deposition(LPCVD). The photocatalytic reaction was examined by measuring the decomposition of benzene gas by gas chromatography under ultraviolet, visible light (λ> 410 nm) irradiation, as well as in the dark. The LPCVD TiO_2-coated $Sr_4Al_{14}O_{25}$:Eu^{2+},Dy^{3+} showed an active photocatalytic reaction under visible irradiation. The mechanism of the photocatalytic reactivity for the TiO_2-coated $Sr_4Al_{14}O_{25}$:Eu^{2+},Dy^{3+} phosphor was identified in terms of the energy band structure and phosphorescence. The coupling of TiO_2 with the $Sr_4Al_{14}O_{25}$:Eu^{2+},Dy^{3+} phosphor might result in energy band bending in the junction region, which makes the TiO_2 crystal at the interface photo-reactive under visible light irradiation. The LPCVD TiO_2-coated $Sr_4Al_{14}O_{25}$:Eu^{2+},Dy^{3+} phosphor was also photo-reactive in the dark through the light photons emitted from the $Sr_4Al_{14}O_{25}$:Eu^{2+},Dy^{3+} phosphor. In addition, the TiO_2-coated Sr-Al-O long lasting phosphors were characterized by X-ray diffraction, scanning electron microscopy, transmission electron microscopy, Auger electron microscopy, and fluorescence spectrophotometer.

INTRODUCTION

Nanocrystalline titanium dioxide (TiO_2) is used widely as a photocatalyst for a range of applications, such as the elimination of pollutants in air and water, water splitting, self-cleaning and dye-sensitized solar cell. On the other hand, TiO_2 is a wide bandgap semiconductor (3.03 eV for rutile and 3.18 eV for anatase) that can only absorb approximately 3~5% of sunlight in the ultraviolet (UV) region, which greatly limits its practical applications. Considerable research effort has been made to develop new TiO_2-based photocatalysts to extend the light absorption of TiO_2 toward the visible region. The addition of a small amount of noble metals, such as Pt, Rd, Ag, and Au, as dopants was reported to enhance the photocatalytic reactivity of TiO_2 remarkably.[1,2]

The coupling of TiO_2 with other metal oxides, such as SiO_2, SnO_2, WO_3, In_2O_3, (Sr,La)$TiO_{3+\delta}$ and $ZnMn_2O_4$, altered the photocatalytic efficiency and energy range of photoexcitation.[3-5] In addition, the junction of TiO_2 with other semiconductors with a narrow band gap enhances the photocatalytic reactivity under visible light irradiation.[6-7] Implantation with a range of transition-metal ions, such as V, Cr, Mn, Fe, and Ni, accelerated by the high voltage enabled a shift in the absorption band of the TiO_2 catalysts toward the visible light region.[8] In addition, the doping of nitrogen and sulfur into TiO_2 could shift its optical absorption edge from the UV range to the visible light range.[9,10]

The long-lasting characteristics of Eu^{2+}-doped alkaline earth aluminate phosphors have attracted considerable attention for their potential applications in fields, such as luminous paints, safety indicators in emergency cases, electronic instrument dial pads, lighting apparatus and switches, automobile dials and panels, writing and printing inks, and plasma display phosphors.[11] The luminescence is strongly dependent on the host lattice, and can occur from the UV to the red region of the electromagnetic spectrum. Up to now, alkali-earth aluminate phosphors with excellent properties, such as $CaAl_2O_4:(Eu^{2+},Nd^{3+})$ [blue], $Sr_4Al_{14}O_{25}:(Eu^{2+},Dy^{3+})$ [blue-green], $SrAl_2O_4:(Eu^{2+},Dy^{3+})$ [green] and $BaAl_2O_4:(Eu^{2+},Dy^{3+})$ [blue], have been developed for a range of applications.[12,13]

In chemical vapor deposition (CVD), a substrate is exposed to a single or multi-component volatile precursor (in gaseous phase) in an inert atmosphere at a controlled temperature and pressure. These volatile precursors will either react or decompose onto the surface of the substrate, creating the desired, deposited thin-film material. The CVD of TiO_2 is normally carried out through a reaction of titanium tetrachloride ($TiCl_4$) with oxygen, or through a thermal reaction of a titanium alkoxide, such as Ti $(OPr^i)_4$ and TTIP. Djerdja et al.[14] reported nanocrystalline TiO_2 films by CVD on different substrates at a relatively low temperature of 320℃ using $TiCl_4$ as the precursor. They found that the nature of the substrate affected the size and distribution of the nano-grains in the films. The best photocatalytic reaction rate was reported with CVD-deposited TiO_2 films, which have a preferred orientation with a columnar structure for the formation of a larger surface area for a dissociative reaction.[15,16]

In this study, nanocrystalline TiO_2 as thin film was supported on the plate of a long lasting phosphor, $Sr_4Al_{14}O_{25}:Eu^{2+},Dy^{3+}$, coated by LPCVD with titanium(IV) isopropoxide (TTIP, $Ti[OCH(CH_3)_2]_4$) as a precursor and its photocatalytic behavior was examined in terms of the photocatalytic degradation of benzene gas as one of volatile organic compounds (VOCs).

EXPERIMENTAL

Nanocrystalline TiO_2 thin films were deposited on the phosphor plate of $Sr_4Al_{14}O_{25}:Eu^{2+},Dy^{3+}$ using a low-pressure CVD (LPCVD) system. TTIP ($Ti(OC(CH_3)_2)_4$, 99.99%, UP Chem., Korea) and oxygen gas were used as the precursor and reaction gas for deposition, respectively. Argon gas was used as a carrier gas for TTIP at a flow rate of 125 sccm. The LPCVD reactor was constructed as a showerhead type. The liquid precursor was kept at 60℃ in a bubbler placed in a heating jacket to ensure an even temperature distribution. The substrate deposition temperature was 390℃ and the pressure was maintained at 1 torr. The flow rate was controlled using a mass flow controller.

The $Sr_4Al_{14}O_{25}:Eu^{2+},Dy^{3+}$ phosphor was prepared by a conventional sintering process using the following raw materials: $Sr(NO_3)_2$(99%, Aldrich, USA), $Al(NO_3)_3 \cdot 9H_2O$(98%, Kyoto Chem., Japan), Eu_2O_3(99.99%, Aldrich, USA), and Dy_2O_3(99.99%, Aldrich, USA) with a molar ratio of 3.88:14:0.02:0.04. A small amount of B_2O_3 (3wt%) was added to the mixed powders as a flux. The powders were then heated at 1400℃ for 3 hrs in a reducing atmosphere of Ar-5%H_2 gas. The sintering process is described in detail elsewhere.[17]

In order to make the phosphor plate substrate, synthesized phosphor powders were mixed with a liquid sodium silicate ($Na_2SiO_{3-x}H_2O$) by using a stirrer, and then coated repeatedly on the slide glass (76×26×1 mm) up to be 4 mm thickness. The last coating layer surface was covered with phosphor powders only to contact with TiO_2. Finally, the phosphor substrate was dried in a dry oven for 2 hrs at 100℃.

The photocatalytic reaction behavior of the LPCVD TiO_2-coated phosphor was examined by measuring the decomposition of benzene gas using a gas chromatography (GC) system. The photocatalytic reaction chamber was filled with 10 ppm benzene gas and irradiated with UV light [75-W mercury lamp] or a 100 W white light lamp (Sylvania incandescent lamp). In the case of visible light irradiation, a UV-light cut-off lens (Edmund Optics) was inserted just below the white light lamp to filter the UV-light with wavelengths (λ) < 410 nm. After a predetermined irradiation time, the sampling gas was extracted from the reaction chamber using a 1 *ml* syringe for the GC analysis test. The morphology, crystal structure and compositional analysis of the LPCVD TiO_2

and phosphor were characterized by transmission electron microscopy (TEM), X-ray diffraction (XRD), scanning electron microscopy (SEM), energy dispersive spectroscopy (EDS), and Auger electron spectroscopy (AES), respectively.

RESULTS AND DISCUSSION

Two opposite cases in sensor responses were assumed before the mixed gas sensing test. One is that the test sensor shows a gas selective response to other gases, and the other is that no sensor has selectivity to all gases. Initially, gas sensor materials without selectivity might possess multi-functional adsorption sites for both gases during mixed gas flow. In an oxidizing and reducing gas mixture, both gases would adsorb on the surface of the sensing material, and the number of adsorbed gas molecules would be in the same proportion as that of each single gas. If the oxidation effect on the total sensor conductance prevails, a specific amount of oxidizing gas would counterbalance the reducing gases, and the remaining oxidizing species might contribute to the change (augment) in conductance. Therefore, a change in conductance might occur in the form of a subtraction reducing effect from oxidizing reactions. Otherwise, in the case of a mixture with reducing agents, a change in conductance might manifest as the summation of the same type reactions (reducing effects), which can have a synergic effect on the reaction of each species.

Fig. 1 shows the XRD pattern of the LPCVD TiO_2-film deposited on a glass substrate at a deposition temperature of 390°C. The film thickness was approximately 150 nm. A comparison of the observed and standard '*d*' values (JCPDS 21-1272) confirmed that only the anatase-phase was present without a rutile phase. The diffraction peaks were assigned to the (101), (112), (200) and (211) crystal planes of anatase, respectively. A (112)-preferred orientation was reported to enhance the photocatalytic reactivity of anatase-TiO_2.[15]

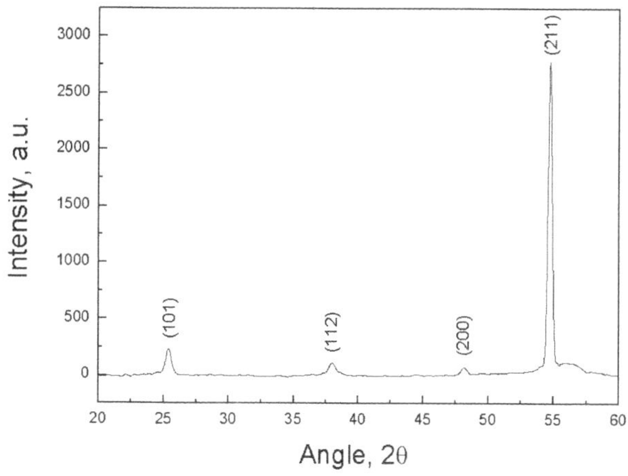

Figure 1. XRD pattern of LPCVD-TiO_2 deposited on the glass at 390℃ for 1min.

Fig. 2 shows (a) a cross-sectional image of TiO_2-coated $Sr_4Al_{14}O_{25}:Eu^{2+},Dy^{3+}$ phosphor and (b) magnified lattice image observed around the interface between TiO_2 and $Sr_4Al_{14}O_{25}:Eu^{2+},Dy^{3+}$ phosphor. As shown in the TEM image, an interface layer of the Sr-Ti-O compounds (mostly $SrTiO_3$) was formed between the TiO_2 and $Sr_4Al_{14}O_{25}:Eu^{2+},Dy^{3+}$ phosphor. The crystallinity of the Sr-Ti-O intermetallic compound is partially amorphous because the lattice image at the interface is similar to a typical amorphous-like image showing unclear lattice fringes compared to the areas of $Sr_4Al_{14}O_{25}:Eu^{2+},Dy^{3+}$ phosphor and TiO_2. This is because TiO_2 was deposited at 390°C, which is less than the temperature for the complete crystallization of Sr-Ti-O compounds, such as $SrTiO_3$, 725°C.[18] Fig. 3 shows the X-ray spectra obtained from energy dispersive spectroscopy (EDS) analysis at three different points marked at the TEM image in Fig. 2(a). The peaks for Sr, Ti, Al and O elements were observed around the interface (point 2) between TiO_2 and $Sr_4Al_{14}O_{25}:Eu^{2+},Dy^{3+}$ phosphor, indicating that the intermixing of Ti, Sr and Al occurred at the interface. Considering the TEM image and EDS spectra, the points marked "1", "2" and "3" at the TEM image correspond to the areas of the $Sr_4Al_{14}O_{25}:Eu^{2+},Dy^{3+}$ phosphor, Sr-Ti-O compounds (mostly $SrTiO_3$), and TiO_2, respectively.

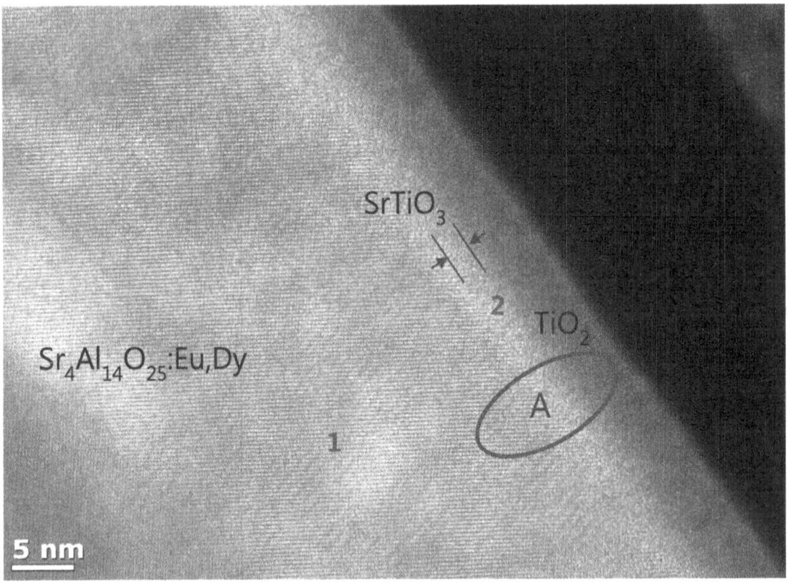

(a)

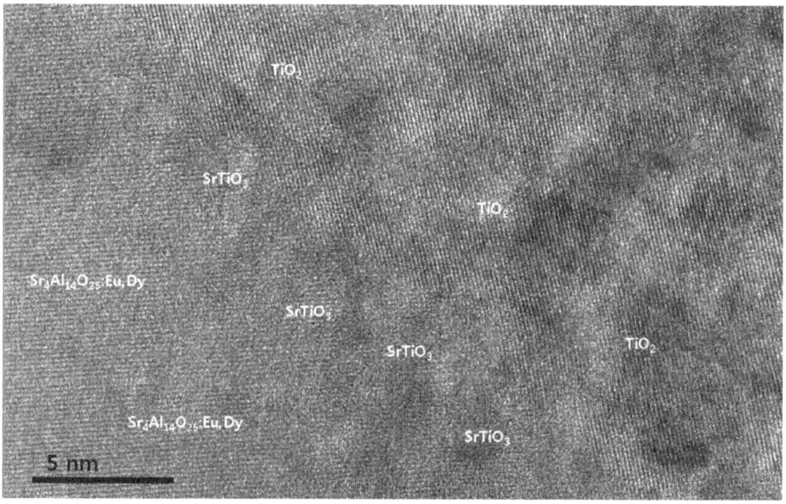

(b)

Figure 2. (a) TEM cross-sectional images of the TiO_2-coated $Sr_4Al_{14}O_{25}$:Eu,Dy phosphor and (b) magnified lattice image at the area of "A" marked in (a).

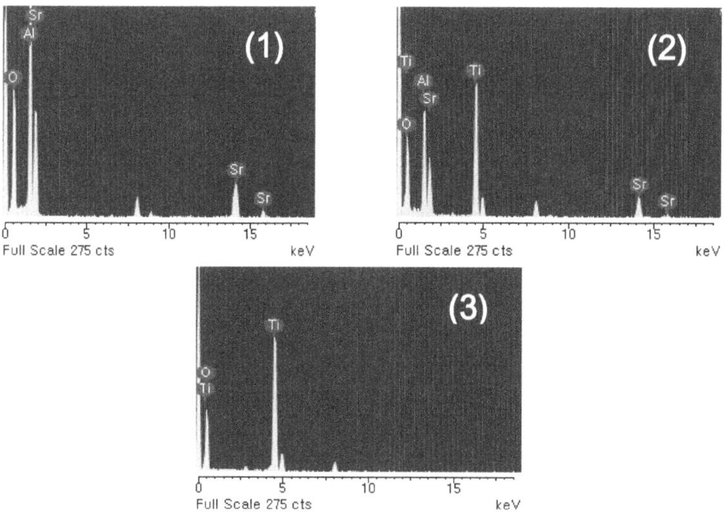

(b)

Figure 3. EDS spectra of the TiO_2-coated $Sr_4Al_{14}O_{25}$:Eu,Dy phosphor at three points marked in the TEM image in Fig.2 (a).

Fig. 4 shows the excitation and emission spectra of the TiO_2-coated phosphor ($Sr_4Al_{14}O_{25}$:Eu^{2+},Dy^{3+}) plate. In addition, the excitation and emission spectra of the pure phosphor was superimposed for comparison. The spectrum shows that excitation occurred through a broad band (over from 250 nm to 450 nm) with the maximum peak at 369 nm and that a broadband greenish-blue emission originating from Eu^{2+} was accompanied with the maximum peak at λ = 493 nm , which is in good agreement with other reports [19,20]. The emission intensity of the TiO_2-coated phosphor was lower than that of the pure phosphor, $Sr_4Al_{14}O_{25}$:Eu^{2+},Dy^{3+}. A TiO_2-coating layer acts as a barrier for the excitation and emission of phosphors, which reduces the emission intensity.

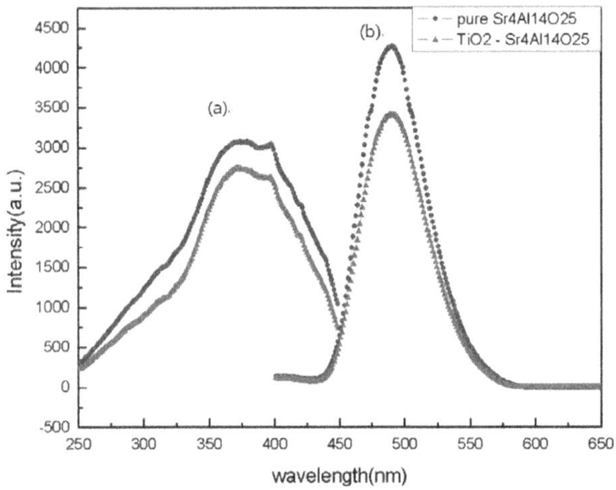

Figure 4. Excitation (a) and emission spectra (b) of the pure phosphor ($Sr_4Al_{14}O_{25}$:Eu^{2+},Dy^{3+}) and TiO_2 coated phosphor.

Fig. 5 shows the changes in the benzene concentration with an irradiation time for the TiO_2 coated phosphor ($Sr_4Al_{14}O_{25}$:Eu^{2+},Dy^{3+}) plates under either UV or visible light irradiation. As shown in the figure for comparison, the benzene concentration for the TiO_2-coated ($Sr_4Al_{14}O_{25}$:Eu^{2+},Dy^{3+}) plates held in the dark state without light irradiation varied slightly with time. On the other hand, the benzene concentration decreased gradually with increasing irradiation time, indicating that the TiO_2-coated phosphor underwent a photocatalytic reaction under both UV and visible light irradiation. The benzene concentration under UV irradiation decreased slightly faster than that under visible light in the first half of the total irradiation time, but was similar to each other in the second half.

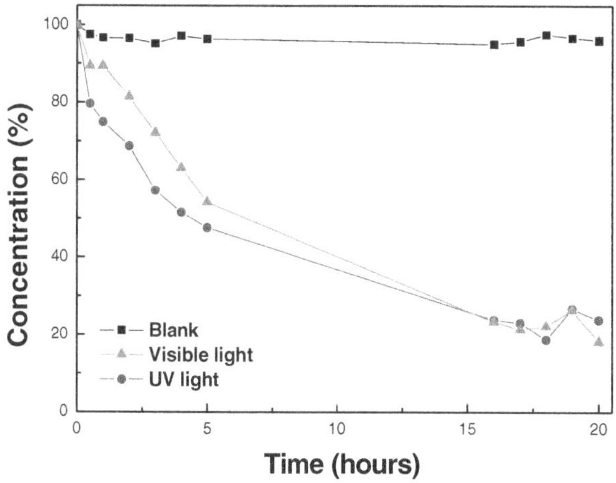

Figure 5. Changes in the benzene concentration with an irradiation time for the TiO_2 coated phosphor ($Sr_4Al_{14}O_{25}:Eu^{2+},Dy^{3+}$) beads under UV and visible light irradiations. The "Blank" symbol stands for the sample of TiO_2-($Sr_4Al_{14}O_{25}:Eu^{2+},Dy^{3+}$) beads held in the dark state without light irradiation.

Pure TiO_2 was photo-excited only under UV-light irradiation. Therefore, the light photons emitted from the ($Sr_4Al_{14}O_{25}:Eu^{2+},Dy^{3+}$) phosphor might not have contributed to the photo-generation of electron-hole pairs because the wavelength was in the range of 410~600 nm. On the other hand, the intermetallic compound of $SrTiO_3$ might have formed through the intermixing of Ti and Sr atoms at the interface between TiO_2 and $Sr_4Al_{14}O_{25}:Eu^{2+},Dy^{3+}$ phosphor. $SrTiO_3$ has a perovskite-type structure with an energy band gap of approximately 3.2 eV,[21] which is similar that of TiO_2 (3.03 eV for rutile and 3.18 eV for anatase).

In the meantime, at the interface layer of $SrTiO_3$, the Sr ion occupies the Ti-site and creates an anion defect as shown in the defect equation using the following Kröger–Vink notation

$$Sr \xrightarrow{Ti} Sr_{Ti}'' + V_O^{\cdot\cdot} \quad \text{--(1)}$$

$$\text{Or} \quad O_O = V_O^{\cdot\cdot} + \frac{1}{2}O_2(g) + 2e' \quad \text{------------------------(2)}$$

where Sr_{Ti}'' and $V_O^{\cdot\cdot}$ are a divalent strontium on a titanium site and an oxygen vacancy, respectively. Referring to the oxygen element of the AES depth profile that decreased at the interface in Fig. 6, the formation of oxygen vacancies was confirmed since oxygen decreased at the interface region between TiO_2 and phosphor ($Sr_4Al_{14}O_{25}:Eu^{2+},Dy^{3+}$). Thus, the defects in the oxygen vacancies occupy the localized levels below the conduction band in the band gap and give extra electrons that are subsequently excited to the conduction band by the photons emitted from the phosphor. The extra excited electrons in the conduction band can participate in a photocatalytic reaction, resulting in an enhancement of the photodecomposition of benzene. The possible energy band structure at the interface between TiO_2 and phosphor is drawn schematically as shown in Fig. 7

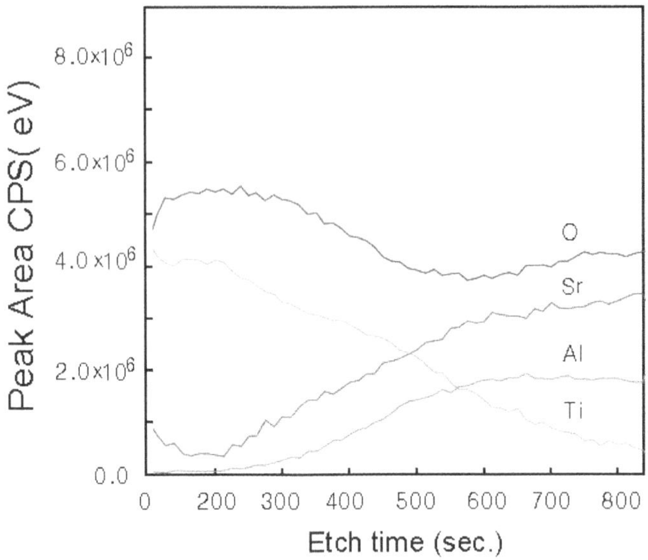

Figure 6. AES depth profiles of the TiO_2-coated phosphor ($Sr_4Al_{14}O_{25}:Eu^{2+},Dy^{3+}$).

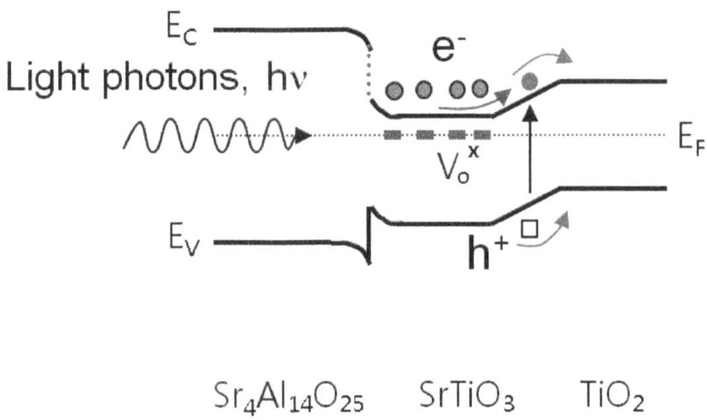

Figure 7. Schematic energy band structure at the interfaces of TiO_2 and phosphor ($Sr_4Al_{14}O_{25}:Eu^{2+},Dy^{3+}$).

In addition, the junction of TiO_2 with $SrTiO_3$ with different Fermi levels (E_F) might induce energy band bending and shift the absorption band of titanium oxide toward the visible light region.[22,23] Therefore, the TiO_2-coated phosphor particles had a larger number of electron-hole pairs, which enhanced the photocatalytic reaction under visible irradiation.

Once the energy bands at the junction area of the TiO_2-$Sr_4Al_{14}O_{25}$:Eu^{2+},Dy^{3+} phosphor are activated by visible light, the emitted light photons from the phosphors enable the photo-generation of electron-hole pairs, engaging the photocalytic degradation of benzene gas. As shown in Fig. 4, the emission intensity of the phosphor was decreased by a coating with TiO_2, but still remained above half the initial intensity, which was sufficient to activate the TiO_2 phase at the interface between TiO_2 and $Sr_4Al_{14}O_{25}$:Eu^{2+},Dy^{3+} phosphor.

Fig. 8 shows the changes in the benzene concentration with time for the TiO_2 coated phosphor ($Sr_4Al_{14}O_{25}$:Eu^{2+},Dy^{3+}) plates under darkness. For comparison, the concentration data obtained for pure TiO_2 (P-25, Degussa) were superimposed in the graph. The samples were irradiated with UV light for 30 min, and then switched-off (dark state). The benzene concentration decreased slightly for pure TiO_2 after the light had been switched-off, indicating that pure TiO_2 was not photo-excited in the dark state. On the other hand, the benzene concentration continued to decrease moderately for the TiO_2-coated $Sr_4Al_{14}O_{25}$:Eu^{2+},Dy^{3+} phosphor, even in the dark state. This suggests that the TiO_2-coated phosphor is photo-reactive under the light photons emitted from the $Sr_4Al_{14}O_{25}$:Eu^{2+},Dy^{3+} phosphor.

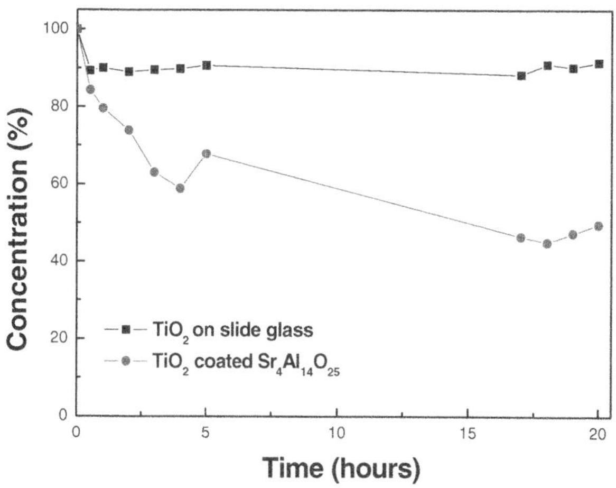

Figure 8. Changes in the benzene concentration with the irradiation time for the TiO_2 coated phosphor ($Sr_4Al_{14}O_{25}$:Eu^{2+},Dy^{3+}) beads and pure TiO_2(P-25) in the dark. The light was switched-off after 30 min irradiation. The thickness of the TiO_2-coated layer was approximately 150 nm.

CONCLUSIONS

The TiO_2 thin film was deposited on the phosphor plate of $Sr_4Al_{14}O_{25}$:Eu^{2+},Dy^{3+} by LPCVD and its photocatalytic activity was investigated by measuring the degradation of benzene gas under either ultraviolet or visible light irradiation, as well as in the dark state. The LPCVD TiO_2-coated $Sr_4Al_{14}O_{25}$:Eu^{2+},Dy^{3+} phosphor showed the photocatalytic degradation of benzene gas under visible light irradiation. The coupling of TiO_2 with $Sr_4Al_{14}O_{25}$:Eu^{2+},Dy^{3+} phosphor might result in energy band bending in the junction region, which makes the TiO_2 crystal at the interface photo-reactive under visible light irradiation. The light photons emitted from the phosphors promoted the photo-generation of electron-hole pairs in the TiO_2-coated $Sr_4Al_{14}O_{25}$:Eu^{2+},Dy^{3+}

phosphor, and enhanced the photocatalytic activity of TiO_2. In addition, the intermetallic compounds of $SrTiO_3$ that are formed at the interface between TiO_2 and the $Sr_4Al_{14}O_{25}:Eu^{2+},Dy^{3+}$ phosphor may result in the formation of oxygen vacancies and additional electrons, subsequently promoting the photodecomposition of benzene gas. The LPCVD TiO_2-coated $Sr_4Al_{14}O_{25}:Eu^{2+},Dy^{3+}$ phosphor was photo-reactive in the dark through the light photons emitted from the $Sr_4Al_{14}O_{25}:Eu^{2+},Dy^{3+}$ phosphor. Benzene gas decomposed continuously on the TiO_2-coated $Sr_4Al_{14}O_{25}:Eu^{2+},Dy^{3+}$ phosphor plate for a considerable amount of time after the light had been switched-off (dark state).

ACKNOWLEDGEMENTS

This research was supported by a grant (14CTAP-C077607-01) from Infrastructure and transportation technology promotion research Program funded by Ministry of Land, Infrastructure and Transport of Korean government.

REFERENCES

[1] Liu, S. X., Qu, Z. P., Han, X. W., & Sun, C. L. (2004). A mechanism for enhanced photocatalytic activity of silver-loaded titanium dioxide. *Catal. Today,* Volume 93-95, 877-884.

[2] Lee, K., Lee, N. H., Shin, S. H., Lee, H. G., & Kim, S. J. (2006). Hydrothermal synthesis and photocatalytic characterizations of transition metals doped nano TiO_2 sols. *Mater. Sci. Eng. B,* Volume 129, 109-115.

[3] Anderson, C. & Bard, A. J. (1997). Improved photocatalytic activity and characterization of mixed TiO_2/SiO_2 and TiO_2/Al_2O_3 materials. *J. Phys. Chem. B,* Volume 101, 2611-2616.

[4] Shchukin, D., Poznyak, S., Kulak, A., & Pichat, P. (2004). TiO_2-In_2O_3 photocatalysts: preparation, characterisations and activity for 2-chlorophenol degradation in water. *J. Photochem. Photobiol. A:Chem.,* Volume 162, 423-430.

[5] Bessekhouad, Y., Robert, D. & Weber, J.-W. (2005). Photocatalytic activity of Cu_2O/TiO_2, Bi_2O_3/TiO_2 and $ZnMn_2O_4/TiO_2$ heterojunctions. *Catal. Today,* Volume 101, 315-321.

[6] Zhu, L., Meng, Z.-D., Cho, K.-Y., & Oh, W.-C. (2012). Synthesis of CdS/CNT-TiO_2 with a high photocatalytic activity in photodegradation of methylene blue. *New Carbon Materials,* Volume 27, 166-174.

[7] Bessekhouad, Y., Brahimi, R., Hamdini, F., & Trari, M. (2012). Cu_2S/TiO_2 heterojunction applied to visible light Orange II degradation. *J. Photochem. Photobiol. A: Chem.,* Volume 248, 15-23.

[8] Yan, M. C., Chen, F., Zhang, J. L., & Anpo, M. (2005). Preparation of controllable crystalline titania and study on the photocatalytic properties. *J. Phys. Chem. B,* Volume 109, 8673-8678.

[9] Asahi, R., Morikawa, T., Ohwaki, T., Aoki, K., & Taga, Y. (2001). Visible-light photocatalysis in nitrogen-doped titanium oxides. *Science,* Volume 293, 269-271.

[10] Sano, T., Mera, N., Kanai, Y., Nishimoto, C., Tsutsui, S., Hirakawa, T., & Negishi, N. (2012). Origin of visible-light activity of N-doped TiO_2 photocatalyst: Behaviors of N and S atoms in a wet N-doping process", *Appl. Catal. B: Environ.,* Volume 128, 77-83.

[11] Blasse G. & Grabmaier B. C. (1994). *Luminescent Materials,* Springer, Berlin, 33-70,.

[12] Nakazawa, E. & Mochida, T. (1997). Traps in $SrAl_2O_4:Eu^{2+}$ phosphor with rare-earth ion doping. *J. Luminescence,* Volume 72–74, 236-237.

[13] Sakai, R., Katsumata, T., Komuro, S. & Morikawa, T. (1999). Effect of composition on the phosphorescence from $BaAl_2O_4$: Eu^{2+}, Dy^{3+} crystals. *J. Luminescence,* Volume 85, 149-154.

[14] Djerdj, I., Tonej, A.M., Bijelić, M., Vranesˇa, V., & Turković, A. (2005). Transmission electron microscopy studies of nanostructured TiO_2 films on various substrates. *Vacuum,* Volume 80, 371-378.
[15] Byun, D., Jin, Y., Kim, B., Lee, J. K., & Park, D. (2000). Photocatalytic TiO_2 deposition by chemical vapor deposition. *J. Hazard. Mater.,* Volume B73, 199-206.
[16] Jung, S.-C., Kim, S.-J., Imaishi, N., & Cho, Y.-I. (2005). Effect of TiO_2 thin film thickness and specific surface area by low-pressure metal–organic chemical vapor deposition on photocatalytic activities. *Appl. Catal. B: Environ.,* Volume 55, 253-257.
[17] Kim, S.-W. & Kim J.-S. (2011). Research on Aftergrow brightness of $Sr_{4-(x+y+z)}Al_{14}O_{25}$:$Eu_x$,$Dy_y$,$Ag_z$ by solid state sythensis. *Kor. J. Met. Mater.,* Volume 49, 348-354.
[18] Sibai, A., Lhostis, S., Rozier, Y., Salicio, O., Amtablian, S., Dubois, C., Legrand, J., Sénateur J. P., Audier, M., Hubert-Pfalzgraff, L., Dubourdieu, C., & Ducroquet, F. (2005). Characterization of crystalline MOCVD $SrTiO_3$ films on SiO_2/Si(100). *Microelectronics Reliability,* Volume 45, 941-944.
[19] Yu, X., Zhou, C., He, X., Peng, Z., & Yang, S. P. (2004). The influence of some processing conditions on luminescence of $SrAl_2O_4$:Eu^{2+} nanoparticles produced by combustion method. *Mater. Lett.* 58, 1087-1091.
[20] Kim, B.-J., Hasan, Z. & Kim, J.-S. (2013). Synthesis and characterization of long persistence $Sr_4Al_{14}O_{25}$:Eu^{2+},Dy^{3+} phosphor prepared by combustion method. *J. Ceram. Proc. Res.* 14(5), 601-605.
[21] Miyauchi, M., Takashio, M., & Tobimatsu, H. (2004). Photocatalytic Activity of $SrTiO_3$ Codoped with Nitrogen and Lanthanum under Visible Light Illumination. *Langmuir,* Volume 20, 232-236.
[22] Otsuka-Yao-Matsuo, S. & Ueda, M. (2004). Visible light-induced photobleaching of methylene blue aqueous solution using $(Sr_{1-x}La_x)TiO_{3+\delta}$-$TiO_2$ composite powder. *J. Photochem. Photobiol. A:Chem.* Volume 168, 1-6.
[23] Yoon, J.-H., Jung, S.-C., & Kim, J. S. (2011). Photocatalytic effects for the TiO2-coated phosphor materials. *Mater. Chem. Phys.,* Volume 125, 342-346.

DEVELOPMENT OF MICROTEXTURED TITANIUM DIOXIDE SURFACE BY USING MICROCUTTING TECHNIQUES

J. Shimizu, T. Yamamoto, L. Zhou, T. Onuki and H. Ojima
Department of Intelligent Systems Engineering, Ibaraki University
Hitachi, Japan

ABSTRACT

TiO_2 possesses photocatalytic functions, even though they are relatively weak and required further enhancements for wider practical usages. On the other hand, improving materials surface functions by introducing surface textures is of great interest in various fields. This study aims to develop a high-functional photocatalytic surface by increasing the real surface area with the support of microcutting techniques. In this work, pure Ti plates, which was covered with numerous fine grooves fabricated by the microcutting and vibration-assisted microcutting techiniques, were anodized to obtain various regularly microtextured TiO_2 film surfaces. Their wettabilties were examined to evaluate their photocatalitic potential. As a result, it is found that the wettability of the TiO_2 film surfaces of regularly textured are superior to that of the mirror finished one under an ultraviolet rays irradiation environment.

INTRODUCTION

It is known that TiO_2 possesses photocatalytic functions such as oxidation-reduction reaction and hydrophilization, when irradiated by ultraviolet rays[1-4]. The expected practical applications of such functions include the deodorization, air/water purification, sterilization, self-cleaning, anti-soiling, anti-fogging and so on. However, further enhancements in their performances are required because their activities are still weak. Hence, a great research effort has been directed towards improving photocatalytic performance by introducing porous film thinning, atomization, surface refinement, visible light responsiveness and so on.

On the other hand, improving materials surface functions by introducing nano/micro surface textures mimicking skins of some plants or creatures is of great interest in various fields. For example, the moth-eye has an antireflective surface. Its optical properties have been studied[5] and concept has also been applied to the commercial products such as the solar cells. In such cases, regularly textured surface as that of the moth-eye is difficult to be obtained and an expansion to various materials is also difficult, since the fabrication method is based on the chemical etching of silicon.

Under such circumstances, this study aims to develop a high-functional photocatalytic film, of which surface is regularly textured by the microscratching/cutting technique to earn the real surface area[6]. In this report, a regularly microtextured TiO_2 film surfaces were fabricated by conducting the anodic oxidation on the titanium plates whose surface was textured by the microcutting and vibration-assisted microcutting techniques[7], and their wettabilty was investigated for evaluating their potential as the photocatalytic surfaces.

PHOTOCATALYTIC REACTION

Figure 1 shows the schematic illustration of the photocatalytic reaction of TiO_2 film in the air. It is considered that O_2 contacted with the TiO_2 film is firstly transformed into O_2^-. O_2^- is then resolved into O^- due to the 'electron or hole effect' promoted by the irradiation of ultraviolet rays. O^- is further transformed into O_3^- when reacting with O_2 in the air. The three types of active oxygen, electron and hole are important factors that influence the photocatalytic reaction, among which the influence of hole is considered to be the most important[8,9], although its mechanism has not yet been fully understood.

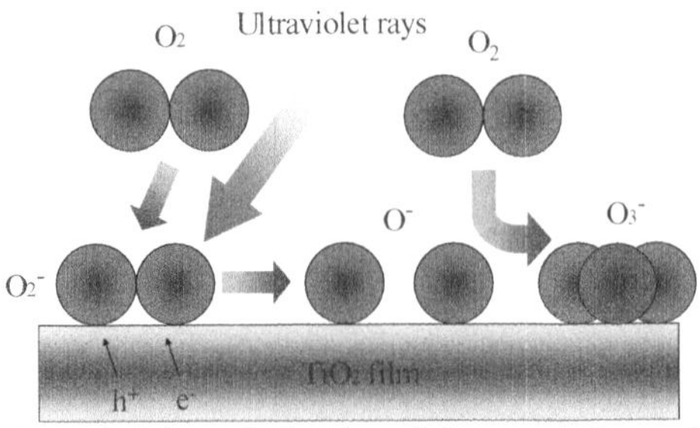

Figure 1. Schematic drawing of photocatalytic reaction of TiO_2 film.

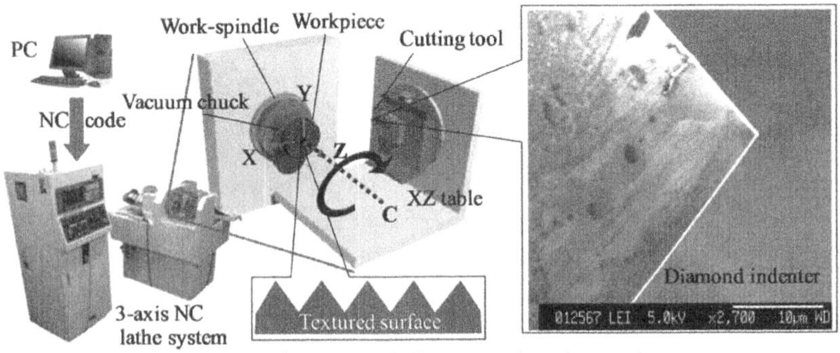

Figure 2. Schematic drawing of 3-axis NC lathe system for microcutting grooves texture.

Table 1: Microcutting conditions for surface texturing.

Workpiece	Pure Titanium
Cutting tool	Monocrystalline diamond
Three edge angle (°)	90 (Triangular indenter)
Edge radius (nm)	500 (initially 100)
Rake angle (°)	-35
Tool rake direction	Face
Depth of cut (um)	9
Feed rate (um/rev)	9, 18, 27
Cutting speed (m/min)	140 (in average)
Environment	Dry (in air)

EXPERIMENTAL

Microcutting for Surface Texturing

Figure 2 illustrates a three-axial (x, y, z) simultaneous NC control machine tool system for the microgrooves formation on Ti plates by using the microcutting technique. A triangular pyramidal indenter for hardness tester is used as a cutting tool, to generate grooves at micrometer scale. Its tip radius and three edge angle are 500 nm (initially 100 nm) and 90°, respectively. A

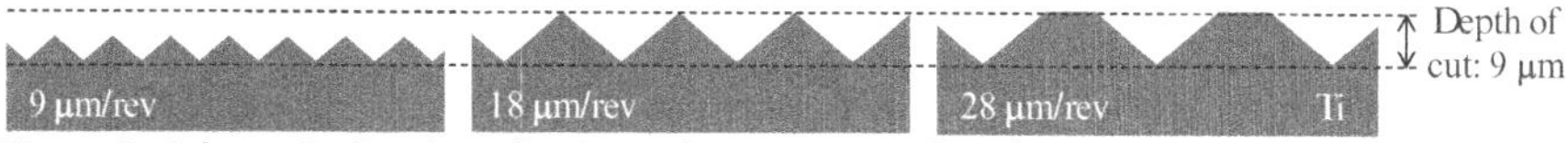

Figure 3. Schematic drawing of estimated cross-sectional surface geometries for three different kinds of feed rates in microcutting.

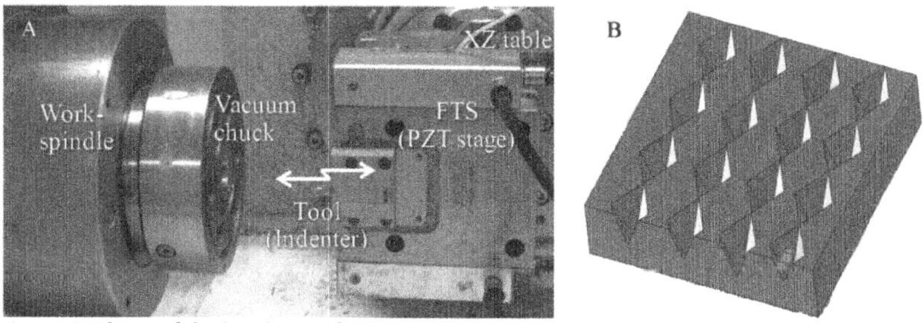

Figure 4. Photo of the head parts for vibration-assisted microcutting (A) and schematic drawing of fabricated surface texture (B).

Table 2: Vibration-assisted microcutting conditions for surface texturing.

Workpiece	Pure Titanium
Cutting tool	Monocrystalline diamond
Three edge angle (°)	90 (Triangular indenter)
Edge radius (nm)	800 (initially 100)
Rake angle (°)	-35
Tool rake direction	Edge
Average depth of cut (um)	6
Feed rate (um/rev)	15
Cutting speed (mm/min)	318 (in average)
Vibration frequency (Hz)	900
Vibration amplitude (um)	6
Environment	Mineral oil

series of the microcutting conditions are listed in Table 1. Figure 3 shows the schematic drawing of estimated cross-sectional surface geometries for three different kinds of feed rates listed in Table 1.

Vibration-assisted Microcutting for Surface Texturing

On the other hand, in the case of the surface texturing by vibration-assisted microcutting, the cutting tool was vibrated in the cutting depth direction by using a FTS (fast tool servo) system, which is a kind of the PZT actuator to realize more complicated tool trajectories than the microcutting mentioned above. A series of the vibration-assisted microcutting conditions are listed in Table 2. A similar cutting tool as the microcutting was used here. Its initial tip radius was 100 nm, however, actually utilized tip radius was 800 nm, because it was a little bit worn during the former machining including the preparation of the initial mirror surface by the

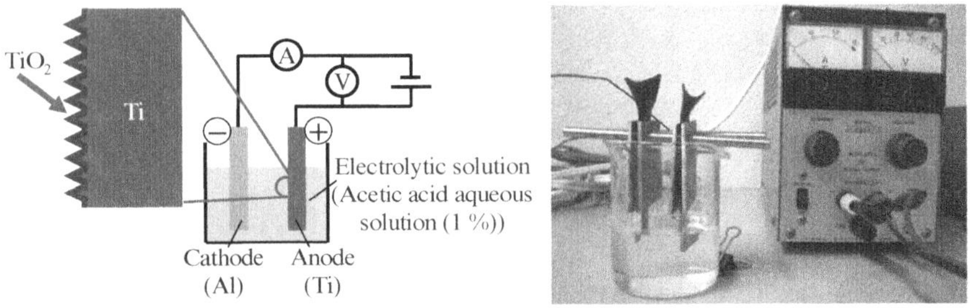

Figure 5. Schematic drawing and outlook of equipment for anodic oxidation.

Table 3: Anodic oxidation conditions for TiO_2 formation.

Anode (Workpiece)	Pure Titanium
Cathode	Aluminum alloy
Electrolytic solution	1% acetic acid aqueous solution
Applied voltage (V)	80
Duration time (h)	1
Distance between electrodes (mm)	15

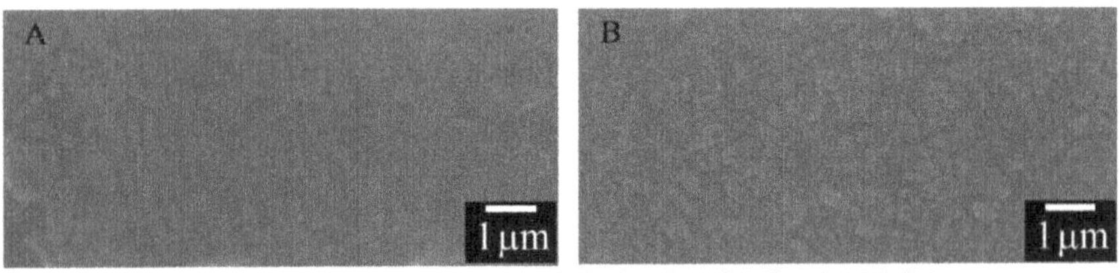

Figure 6. SEM images of polished Ti (A) and polished and oxidized Ti (TiO_2) (B).

microcutting. The vibration frequency and amplitude are set to 900 Hz and 6 um, respectively, as the first trial. A much lower cutting speed than the microcutting is applied here to fabricate numerous numbers of periodical micro-patterns in the cutting direction.

Anodic Oxidation for TiO_2 Formation

Figure 5 shows the schematic drawing and outlook of equipment for anodic oxidation experiments. The anodic oxidation conditions for TiO_2 formation are listed in Table 3. A pure Ti plate and an Al alloy plate are used for the anode and cathode respectively, and 1% acetic acid aqueous solution, which is one of the environmentally-friendly solutions, is used as the electrolytic solution. The applied voltage is set at 80 V referring to the preferable condition for the anatase structure formation of TiO_2.

SEM micrographs of polished Ti (A) and polished and anodic oxidized Ti (TiO_2) (B) surfaces are shown in Fig. 6. The generation of countless numbers of orange peel structures several hundreds of nm in size can be observed. This is preferable since it increases the real surface area. In this case, about 40 - 60 nm thick TiO_2 layer was generated on the anode surface, since the color of the surface turned light blue.

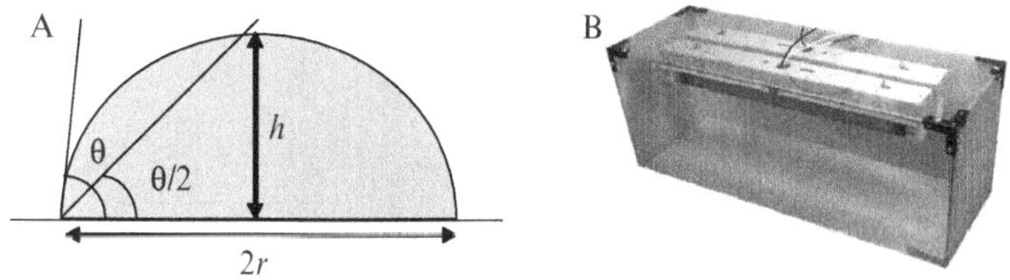

Figure 7. Illustration of half-angle method (A) and outlook of wettability testing apparatus (B).

Table 4: Conditions for wettability tests.

Amount of water droplet (ul)	4
Light wavelength (nm)	300 - 400 (peak: 350)
Temperature (°C)	25 (Microcutting), 26 (Vibration-assisted microcutting)
Relative humidity	30 (Microcutting), 24 (Vibration-assisted microcutting)

Evaluation of Wettability

The influence of surface area of TiO_2 film on the hydrophilicity was evaluated, and the half-angle method was used for the determination of the contact angle theta. Figure 7 shows illustration of the half-angle method (B) and outlook of equipment for the wettability test under an ultraviolet rays irradiation environment (A), respectively.

The wettability tests were conducted on fabricated TiO_2 surfaces with 4 ul purified water droplets by using a testing apparatus shown in Fig. 7(B). The apparatus is equipped with a pair of 15 W black light sources whose wavelength is 300 - 400 nm (peak is 350 nm) and covered by acrylic plates to prevent the influence of wind on the measurement results.

The conditions for wettability tests are listed in Table 4.

RESULTS AND DISCUSSION

Surface Texture by Microcutting Technique

Figure 8 shows the micrographs of textured pure Ti by microcutting under three different kinds of feed rates 9 um/rev (A), 18 um/rev (B) and 27 um/rev (C), and oxidized surface of (B) (D), respectively. From Figs. (A) to (C), it can be clearly observed that the pitch of the cutting groove in the horizontal direction increases with the feed rate and burr is generated in between each two neighbor cutting grooves. Such burrs are preferable to increase the real surface area and to enhance the photocatalitic reactions. The oxidized surface in Fig. 8(D) is a little rougher than Fig. 8(B) due to the formation of orange peer structures mentioned above.

In order to quantitatively evaluate the variation in real surface area due to texturing, the surface area ratio Ar/Aa was examined, where Ar is the real surface area measured by a surface tester and Aa is the appearance surface area. Figure 9(A) shows the relationship between surface area ratio and feed rate. From Fig. 9(A), it can be understood that the minimum feed rate of 9 um/rev shows almost the maximum surface area ratio 1.44 and the reduction in feed rate is advantageous to increase the real surface area. Fig. 9(A) indicates that the anodic oxidation is also advantageous for enlarging the real surface area due to the generation of aforementioned orange peer structures.

In order to evaluate the photocatalitic function in the textured TiO_2 surfaces, wettability tests were conducted. In this case, the wettability in the cutting direction was examined. Figure 9(B) shows the relationship between contact angle of water droplet and duration of UV

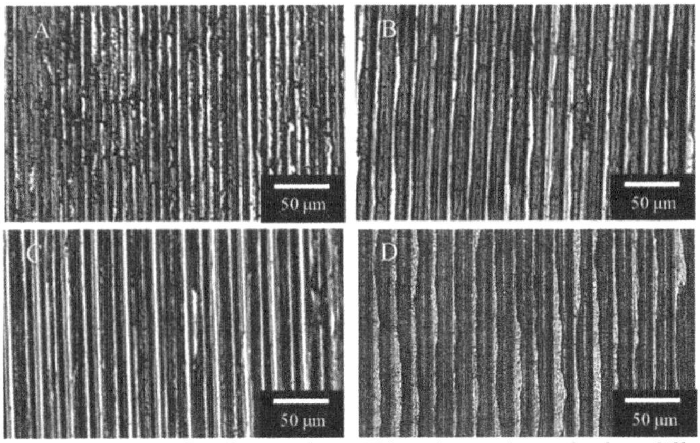

Figure 8. Micrographs of textured pure Ti by microcutting under three different kinds of feed rates 9 um/rev (A), 18 um/rev (B) and 27 um/rev (C), and oxidized surface of (B) (D).

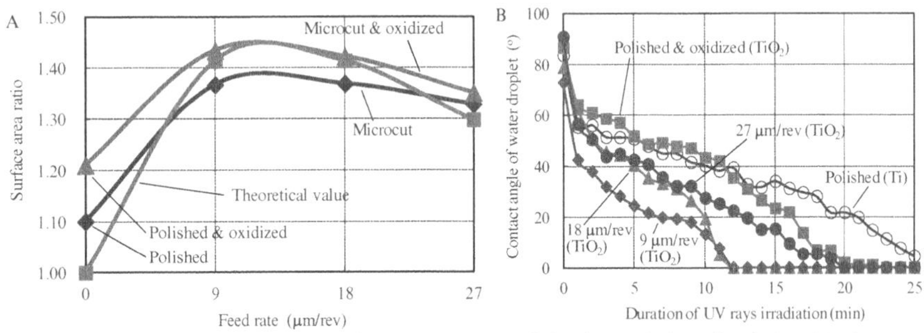

Figure 9. Relationship between surface area ratio and feed rate (A) and relationship between contact angle and duration of UV rays irradiation (B).

rays irradiation. By comparing Figs. 9(A) and (B), it is figured out that the wetting speed (decreasing speed in contact angle) increases with the increase in surface area ratio, and it marks the maximum when the feed rate is 9 um/rev.

Surface Texture by Vibration-assisted Microcutting Technique

Figure 10 shows the micrograph of textured pure Ti by vibration-assisted microcutting (A) and 3-D image of its oxidized surface (B) measured by a laser microscope, respectively. From Fig. 10, it is observed that the microgrooves reflecting the cutting tool vibration as well as feed motion are regularly formed on the surfaces. They are covered with numerous micro impressions and pile-ups. The patterns pitches in the feed and cutting directions are 15 um and 7.3 um, respectively. The small pile-ups or burrs can be seen around each groove, and their maximum height becomes about 2.5 um. Such pile-ups or burrs formations are preferable to enlarge the real surface area and to activate the photocatalytic reaction.

A comparison result of the surface area ratio is shown in Fig. 11(A). The measurements were done before and after the anodic oxidation and the surface area ratio is standardized using

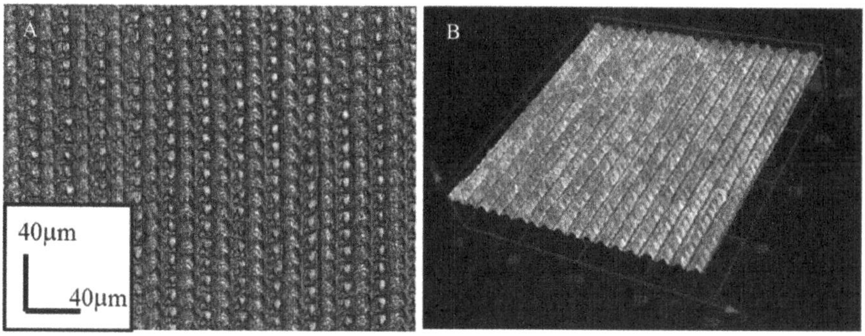

Figure 10. Micrograph of textured pure Ti by vibration-assisted microcutting (A) and 3-D image of its oxidized surface (B).

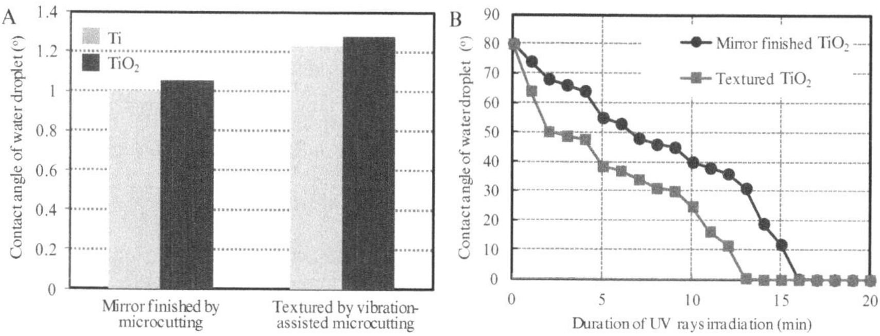

Figure 11. Comparison result of surface area ratio between mirror finished and textured surfaces (A) and relationship between contact angle and duration of UV rays irradiation (B).

the value of the mirror finished one here. The real surface area was calculated by the laser microscope. However, the comparison between the two different kinds of microcutting methods were not done here, since the fabrication conditions were a little different each other. From Fig. 11(A), it is found that the surface area ratio of the microtextured and anodized surface becomes 1.28, which is 1.22 times as large as that of mirror finished and anodized one (1.05).

Figure 11(B) shows the relationship between contact angle of water droplet and duration of UV rays irradiation on fabricated TiO_2 film surfaces. It is found that the time taken to be 0° in contact angle in the case of the microtextured surface (13 min) is 3 minutes shorter than that by mirror finished one (16 min). By comparing Figs. 11(A) and (B), it can be also recognized that the wetting speed increases with the surface area ratio.

From these results, it is proven that the high-performance TiO_2 films can be fabricated by the combination of the microcutting and the anodic oxidation techniques, even though the more precise evaluations on the relationship between surface topography related with the fabrication method and photocatalitic functions will be required.

CONCLUSIONS

In order to develop a high-performance photocatalytic surface, the surface texturing by microcutting and vibration-assisted microcutting of pure Ti was performed. Then, textured Ti plates were anodized in order to generate the TiO_2 films on the uppermost surface of the textured Ti plates. The wettabilty, which is one of the photocatalytic functions, of the textured TiO_2 film surfaces were evaluated. As a result, it is found that the wettability of TiO_2 film surfaces regularly microtextured are superior to that of the mirror finished one due to the increase in real surface area, and numerous micro burrs and pile-ups incidentally formed by the microcutting techniques are also advantageous to increase the real surface area and enhance the photocatalytic reactions.

ACKNOWLEDGEMENTS

This work was partially supported by Grant-in-Aid for Fundamental Research (C) (254200432 and 15K05753) from Japan Society for the Promotion of Science.

REFERENCES

[1] Fujishima, A., & Honda, K. (1972). Electrochemical Photolysis of Water at a Semiconductor Electrode. *Nature, Volume 238*, 37-38.

[2] Wang, R., Hashimoto, K., Fujishima, A., Chikuni, M., Kojima, E., Kitamura, A., Shimohigoshi, M., & Watanabe, T. (1997). Light-Induced Amphiphilic Surface. *Nature, Volume 388*, 431-432.

[3] Miyauchi, M., Nakajima, A., Fujishima, A., Hashimoto, K., & Watanabe, T. (2000). Photo-induced surface reactions on TiO_2 and $SrTiO_3$ films: Photocatalytic Oxidation and Photo-induced Hydrophilicity. *Chem. Mater., Volume 12*, 3-5.

[4] Sun, R., Nakajima, A., Fujishima, A., Watanabe, T., & Hashimoto, K. (2001). Photoinduced Surface Wettability Conversion of ZnO and TiO_2 Thin Films. *J. Phys. Chem. B, Volume 105*, 1984-1990.

[5] Wilson, S. J., & Hutley, M. C. (1982). The Optical Properties of 'Moth eye' Antireflection Surfaces. *Optica Acta, Volume 29*, 993-1009.

[6] Shimizu, J., Zhou, L., Takamori, K., Ojima, H., Yamamoto, T., & Huang, H. (2010). Enhancement of Photocatalytic Reaction of Titanium Dioxide Film by Surface Texturing. *Materials Science Forum, Volume 654-656*, 1784-1789.

[7] Shimizu, J., Yamamoto, T., Zhou, L., Ojima, H., Onuki, T., & Nagaoka, S. (2013). Fabrication of Surface Microtexture by Vibration Assisted Cutting. *Advanced Materials Research, Volume 797*, 638-641.

[8] Sakai, N., Fujishima, A., Watanabe, T., & Hashimoto, K. (2001). Enhancement of the Photoinduced Hydrophilic Conversion Rate of TiO_2 Film Electrode Surfaces by Anodic Polarization. *J. Phys. Chem. B, Volume 105*, 3023-3026.

[9] Sakai, N., Fujishima, A., Watanabe, T., & Hashimoto, K. (2003). Quantitative Evaluation of the Photoinduced Hydrophilic Conversion Properties of TiO_2 Thin Film Surfaces by the Reciprocal of Contact Angle. *J. Phys. Chem. B, Volume 107*, 1028-1035.

MORPHOLOGY CONTROL AND PHOTOCATALYTIC ACTIVITY OF TiO_2 FILM

Jinshu Wang, Hongyi Li, Junshu Wu, Qian Cai, Yilong Yang, and Bingxin Zhao

School of Materials Science and Engineering, Beijing University of Technology, Chaoyang District, Beijing 100124, China
Corresponding author e-mail: wangjsh@bjut.edu.cn

ABSTRACT

Using anodized pure aluminum, 3102 aluminum alloy and 6061 aluminum alloy as the templates, TiO_2 thin films were prepared through a liquid deposition method. The TiO_2 nanotubes fabricated on the 3102 aluminum alloy substrate had one-dimensional tubular structures, similar to that obtained on pure aluminum substrate, whereas that prepared on the 6061 aluminum alloy substrate had bamboo-like structure with holes. It was found that the growth of TiO_2 nanotubes or nanorods was correlated with AAO templates. The morphology control and morphology transformation mechanism were discussed. The results showed that the deposition temperature in the range of 30-40°C was favorable for the formation of TiO_2 nanoarrays. The photocatalytic capability of the TiO_2 films prepared on the aluminum alloy is superior to that of the films fabricated on the pure aluminum. The main alloying elements Si was doped into TiO_2 lattice to form Ti-O-Si bonds, resulting in the enhanced photocatalytic activity.

INTRODUCTION

Photoelectrochemical processes at semiconductor colloid-electrolyte interfaces have been received special attention because of their possible applications in the conversion of solar energy into chemical energy and pollution control[1-6]. Photocatalytic reaction systems have been commercially supplied to conduct self-cleaning, deodorant, anti-bacteria, etc., and are expected to be applied to the pollution control to decompose toxic materials in air and waste water using sun light and indoor light. Among all the photocatalysts, only TiO_2 has been commercially used because of the excellent photocatalytic activity and chemical stability. In view points of its utility in such a field, compared with TiO_2 powder, processing TiO_2 as thin films or coatings with high crystallinity, high specific surface area is favorable for easy separation, recovery and high recycle rate. Crystalline TiO_2 has many morphologies such as nanofibers, nanoparticles, nanorods, nanospheres, nanotubes and nanowires[7-9]. TiO_2 nanotubes are highly efficient in photocatalysis since they have a relatively higher interfacial charge transfer rate and surface area compared with spherical TiO_2 particles[10].

Many approaches have been developed for the preparation of TiO_2 nanotubes, i.e., chemical vapor deposition (CVD), anodic oxidation, seeded growth, the wet chemical (hydrothermal method, the sol-gel method) and liquid phase deposition (LPD) of templates[11-17]. Among all these approaches, LPD of template method is one of the simplest and most practical one to fabricate TiO_2 nanotubes, since it has so many advantages such as low cost, mild reaction condition, simple equipment requirement and allows TiO_2 films to be deposited over large areas. Anodic aluminum oxide (AAO) is one of the commonly used templates due to its unique structure, such as controllable pore diameter, extremely narrow pore size distribution, and ideally cylindrical pore shape[14]. TiO_2 nanohole arrays by LPD method using anodic alumina disk have been reported[15-17]. However, the diameter of AAO templates prepared with different electrolytes, the corresponding morphology of TiO_2 nanoarrays and the effect of aluminum alloy substrate on the morphology of AAO templates, have seldom been systematically studied. In this work, we studied the morphology control of TiO_2 nanoarrays prepared with different AAO templates and discussed morphology change mechanism. Dyes are important industrial pollutants and methyl blue is famous for its good stability as dye materials. The degradation of methyl blue aqueous solution is usually used to evaluate the performances of photocatalysts under UV and visible light irradiation[18-20]. Therefore, the photodegradation of methyl blue by the TiO_2 nanotube films on pure aluminum substrate and aluminum alloy substrates have been studied.

EXPERIMENTAL

PREPARATION OF AAO TEMPLATES

A high purity annealed aluminum foil (99.96% purity), 3102 alloy (98.05% purity) and 6061 alloy (97% purity) foils were used as starting materials. The rough composition of 3102 alloy was a minimum of 98.05 wt.%, with Fe< 0.7 wt.% and Si< 0.4wt.% as principal alloying elements. The rough composition of 6061 alloy was a minimum of 97 wt.% Al, with Mg<1.2 wt.% and Si<0.8 wt.% as principal alloying elements. The foils for anodizing were cut into pieces with dimensions of 30mm×40mm. Prior to anodization, organic impurities and the oxide layer on the foil surface were removed by ultrasonic washing in deionized water, ethanol, and acetone in sequence. The pure aluminum foils were anodized in aqueous solution of oxalic acid, sulfuric acid and phosphoric acid or mixed electrolytes, respectively. The other two aluminum alloy foils were anodized in aqueous solution of phosphoric acid.

PREPARATION OF TIO_2 NANOTUBE

Then the anodic alumina templates were immersed in 0.1mol/L $(NH_4)_2TiF_6$ for 20min. With the aqueous flowing along the wall of the AAO templates, TiO_2 nanoarrays could be in-situ formed in the inner side of AAO template. The deposited TiO_2 nanofilms were annealed at 550°C for 2h.

PHOTOCATALYTIC ACTIVITY MEASUREMENT

The photocatalytic activities of the samples were measured by degradation of methyl blue (MB). The samples were immersed in 100mg/L methyl blue solution for 24h in the dark. Then the samples were removed from the solution with the methyl blue on the surface. Under the Hg lamp irradiation (CMH-250, 10mW/cm^2), the degradation of methyl blue on samples' surface began. Methyl blue decomposition was estimated by the degradation fraction (Eq.1), where the wavelength of 580 nm was monitored by UV–Vis spectrometer (UV2550).

$$\eta = \frac{REF_t - REF_{initial}}{REF_0 - REF_{initial}} \quad (1)$$

$REF_{initial}$: the reflectivity of the samples with adsorption of methyl blue without light irradiation;
REF_t: the reflectivity of the samples with adsorption of methyl blue under light irradiation for various time;

REF_o: the reflectivity of the samples without adsorption of methyl blue.

RESULTS AND DISCUSSION

CHARACTERIZATION OF AAO TEMPLATES

In order to get AAO templates with good morphology, pure aluminum foils were used and anodized using different electrolytes. Figure 1 shows the morphology of AAO templates prepared with oxalic acid, sulfuric acid and phosphoric acid or mixed electrolytes. All the AAO templates had hexagonal shape which was caused by high electric field squeezing[21]. The outside diameters of AAO obtained in different electrolytes were in the order: AAO prepared with phosphoric acid > That with mixed electrolytes of oxalic acid and phosphoric acid > That with oxalic acid > That with mixed electrolytes of sulfuric acid and phosphoric acid ≈ That with sulfuric acid. The AAO prepared with phosphoric acid had the largest pore size, i.e., the mean outside diameter of the tubule and inside diameter were about 250 nm and 150 nm, respectively, as shown in Fig. 1(a). On the other hand, the AAO prepared with sulfuric acid had the smallest pore size (Fig.1(d)), i.e., the mean outside diameter and inside diameter of that prepared with oxalic acid were about 50 nm and 20 nm, respectively. The mean diameter of the AAO templates prepared with mixed electrolytes of oxalic acid and phosphoric acid was in the middle with the outside diameter of 200 nm and inside diameter of 100 nm, as shown in Fig.1(c). Figure 1 (e) also showed that although the AAO templates prepared with mixed electrolytes of phosphoric acid and sulfuric acid had smaller pore size, the pores existed orderly and uniformly. The results indicated that the pore size was correlated with the kind of electrolyte. Sulfuric acid had good dissolving capability, therefore, under the lower voltage, Al which located at underneath layer dissolved first and the initial cavity could be formed in the oxide film. Phosphoric acid had a weak oxidizing ability, therefore, Al could only be dissolved under higher voltage. It was reported that the highest anodizing voltage applied in the electrolyte was in the order of phosphoric acid> oxalic acid [22]. It could be deduced that the highest anodizing voltage applied in the electrolyte of phosphoric acid was the largest among those applied in all five electrolytes. To obtain the templates with large pores, high anodizing voltage should be applied. Therefore, in the following anodizing process for aluminum alloys, phosphoric acid was chosen as the electrolyte.

Figure 2 illustrates the side view of the AAO templates prepared with pure aluminum, 3102 aluminum alloy and 6061 aluminum alloy foils. It could be seen that the AAO templates prepared with 6061 aluminum alloy appeared as bamboo-like structure with holes (Fig.2(c)), whereas those prepared with pure aluminum (Fig.2(a)) and 3102 aluminum alloy (Fig.2(b)) had one-dimensional columnar structures. The difference in the AAO morphology might be caused by the alloying elements which took different oxidation behavior during the anodizing process. Gibbs free energy per equivalent, $\Delta G_0/n$ for the formation of the alloying element oxide, and $(\Delta G_0/n)_{Al}$ for the formation of Al_2O_3, was different. When $\Delta G_o/n$ of the alloying oxide was smaller than $(\Delta G_o/n)_{Al}$, the alloying element could be easily oxidized, therefore, alloying element oxide could be formed during the anodizing process. Otherwise, Al would be firstly oxidized and the alloying elements existed in the films as its original state[23]. The main elements in 6061 aluminum alloy are Mg(0.8%~1.2wt.%), Si(0.4%~0.8wt.%), and those in 3102 aluminum alloy are Fe(<0.7wt.%), Si(<0.4wt.%). Based on the $\Delta_f G^{\ominus}_m$ of Al_2O_3 (-1582.3kJ/mol

[24]), the calculated $(\Delta G_o/n)_{Al}$ was -263.72 kJ/mol. As for the other three elements, the calculated $(\Delta G_o/n)_{Mg}$, $(\Delta G_o/n)_{Si}$ and $(\Delta G_o/n)_{Fe}$ were -284.71 kJ/mol, -212.67kJ/mol and -123.7KJ/mol, respectively. Therefore, the $(\Delta G_o/n)$ of these four elements were in the order $(\Delta G_o/n)_{Mg}<(\Delta Go/n)_{Al}< (\Delta G_o/n)_{Si}< (\Delta G_o/n)_{Fe}$. In other words, apart from the oxidation of Al, Mg could be oxidized into MgO, whereas Si and Fe kept their original state. PBR[25], the volume ratio of metal oxide and pure metal, could be applied for describing the morphology difference of the oxide layer. The PBR of MgO/Mg is smaller than that of Al_2O_3/Al, leading to the formation of holes in the oxide layer, as shown in Fig. 2(c). Therefore, the alloying elements resulted in the irregular growth of oxide film at the surface of 6061 aluminum alloy, while the growth of oxide film on 3102 aluminum alloy were not affected by the alloying elements.

CHARACTERIZATION OF TIO_2 NANOARRAYS

By immerging the AAO templates on pure aluminum substrates into $(NH_4)_2TiF_6$ solution, TiO_2 nanoarray films were prepared and the morphologies were shown in Fig. 3. As shown in Fig. 3(a), TiO_2 prepared with AAO template using oxalic acid exhibited as the nanorods with the mean diameter about 100 nm. The uniform and well aligned TiO_2 nanorods were perpendicular to aluminum substrates. Furthermore, this structure was similar to that of AAO as comparing Fig. 3 (a) to Fig. 1(b). The pore diameter of AAO template was also about 100nm, indicating that hydrous TiO_2 deposited in-situ on anodic alumina template and filled in the tubule. However, TiO_2 prepared with the AAO template using sulfuric acid was covered by a thick deposited substance on the surface after LPD process. After ultrasonic treatment, it was clear to see in Fig.3 (c) that TiO_2 existed as naonorods with the diameter of 50 nm, whereas using the AAO templates obtained with phosphoric acid, mixed electrolytes of phosphoric acid and oxalic acid, or mixed electrolytes of phosphoric acid and sulfuric acid, all TiO_2 existed as nanotubes, but the nanotube diameter varied with the electrolyte, which were summarized in Table 1. TiO_2 prepared with AAO obtained with phosphoric acid had the largest diameter with the outside diameter of 200 nm and inside diameter of 180 nm, as displayed in Fig. 3(d), since the pore size of AAO templates obtained with phosphoric acid was the largest among all the electrolytes. By comparing the morphology of TiO_2 nanotube obtained with different electrolytes, TiO_2 nanotubes prepared with the mixed electrolytes of sulfuric acid and phosphoric acid had ordered and uniform structure with the smaller outside diameter of 50 nm, as shown in Fig.3(e). From Figure 3, it was evident that the formation of TiO_2 nanorod or nanotube was correlated with the pore size of the AAO templates.

The formation mechanism of nanotube and nanorod could be discussed based on the formation of hydrous TiO_2 as illustrated in reactions (2) and (3) .

$$[TiF_6]^{2-}+6H_2O \leftrightarrow [Ti\,(OH)_6]^{2-}+6H^++6F^- \quad (2)$$

$$Al_2O_3+12H^++12F^- \leftrightarrow 2H_3AlF_6+3H_2O \quad (3)$$

The above chemical reactions indicated that the production of 2 mol TiO_2 needed to consume 1 mol Al_2O_3. TiO_2 prepared after LPD existed as hydrous titania and would transform into anatase phase after heat treatment. Al_2O_3 in anodic alumina template had the phase of γ-Al_2O_3[26]. It could be calculated that 1 mol Al_2O_3 with the volume of 29.14 cm^3 would be consumed for the formation of 2 mol TiO_2 with the volume of 41.67 cm^3 based on the density of γ- Al_2O_3 (3.5 g/cm^3) [26] and anatase (3.84 g/cm^3)[27]. So the volume ratio of $V_{Al_2O_3}/V_{TiO_2}$ was 0.71, indicating that TiO_2 produced by the in-situ synthesis would cover all the template surface when the ratio of volume of Al_2O_3 to the total volume of the template ($V_{Al_2O_3}$ and V pores) was equal to 0.71.

Table 1 shows the volume ratios of Al_2O_3 in the AAO templates prepared with oxalic acid and sulfuric acid were larger than 0.71, therefore, TiO_2 existed as nanorod and the excess amount of TiO_2 was dissolved into the solution or existed on the container wall since the volume of TiO_2 exceeded the total volume of the templates. Especially for TiO_2 prepared with AAO templates using sulfuric acid, we found a titanium oxide layer on the film surface which might be caused by the deposition of the excess TiO_2, but this deposited layer could be easily removed by ultrasonic treatment. On the other hand, the volume ratio of Al_2O_3 in the AAO templates prepared with phosphoric acid and mixed electrolytes were smaller than 0.71, therefore, TiO_2 existed as nanotubes since TiO_2 could not cover all the AAO template surface.

Figure 4 displays the morphology of TiO_2 prepared with the AAO templates on pure aluminum substrate obtained with phosphoric acid but deposited at different temperature with the same deposition time of 20 min. As described above, TiO_2 grew in-situ with the depletion of alumina. When the temperature was low, the reaction rate was low, therefore, alumina was left, which is shown in Fig.4(a). When the temperature was increased to 35 °C, it was hard to find the remained alumina and only TiO_2 nanotube was observed, as shown in Fig. 4(b). Increasing the temperature to 40 °C and 45 °C, as shown in Fig. 4(c) and (d), the fast reaction rate led to the production of an excess amount of titanium compound which covered on the surface of TiO_2 nanotube. The higher the temperature, the larger the amount of titanium compound. But the covered compound could be removed from the surface after ultrasonic treatment.

From Figure 4, it was obvious that titanium compound was covered at the surface of TiO_2 nanotube layer at the high LPD temperature. During the experiments, we also found that a large number of bubbles escaping from the liquid phase at the temperatures of 30 °C, 35 °C and 40 °C, but only a small number of bubbles were found when the LPD process was carried out at 45 °C. Figure 5 shows the schematic diagram of LPD taken at the high temperature. During the high temperature LPD process, the chemical reaction took place fiercely and a large amount of $[Ti(OH)_6]^{2-}$ would deposit on the inside surface of the AAO tubules and the surface of the AAO templates, accompanied by the simultaneous decomposition of $(NH_4)^+$ to NH_3 which escaped from the AAO tubules. Since the reaction rate was relatively fast at high temperature, the deposited $[Ti(OH)_6]^{2-}$ which covered at the surface hampered the escaping of NH_3. The gas in the nanotube produced compression pressure to the nanotube, leading to the morphology change from original hexagonal shape to circular shape.

Figure 6 displays side view of TiO_2 nanotubes synthesized in situ on different AAO substrates. It could be seen that the morphology of TiO_2 films was similar to that of AAO templates. TiO_2 films prepared with AAO templates on pure alumina and 3032 alumina alloy had uniform structure(Fig.6(a) and Fig.6(b)) whereas that film prepared with AAO templates on 6061 aluminum alloy had a large number holes, as shown in Fig. 6(c), similar to the AAO templates(Fig. 2(c)).

PHOTOCATALYTIC ACTIVITY OF TIO_2 NANOARRAYS

Figure 7 shows the photocatalytic performance of the catalysts by the degradation of aqueous methyl blue. After 2.5h irradiation, 76.5% and 79.4% MB had been photograded by the TiO_2 nanofilms on the 6061 aluminum alloy and 3102 aluminum alloy, respectively, much higher than the TiO_2 nanofilms prepared on pure aluminum (43.9%) under the same experimental conditions. It should be noticed that the TiO_2 films on the 6061 aluminum alloy and 3102 aluminum alloy had nearly the same photocatalytic activity.

In order to find the reason for the difference in the photocatalytic activity, X-ray Photoelectron Spectroscopy (XPS) analyses were carried out for pure TiO_2 and the TiO_2 film on 6061 aluminum alloy substrate to determine the chemical composition. The recorded spectra of were given in Figure 8. Figure 8(a) illustrates that TiO_2 films on 6061 Al alloy were composed of O, Ti, C and Si elements, and C could be attributed to the adsorbed carbon. As shown in Fig.8(d), the peak around 1303.8eV was detected, which was assigned to the peak of Mg1s in MgO. Two peaks at 530eV and 532eV at O1s core level (Fig.8(c)) were assigned to the Ti-O bond and the residual oxygen-containing group, respectively. Figure 8(b) shows that the binding energy of the Ti 2p3/2 peak prepared on 6061 aluminum alloy was 457.2eV, which was 0.9eV greater than that of pure TiO_2. The decrease of the electron density around Ti atoms was due to the greater electronegativity of Si via O than Ti via O[28], indicating that Ti-O-Si bonds were formed on the surface of TiO_2 films. On the other hand, the binding energy of Si 2p was 101.5eV, as shown in Fig.8(e), smaller than that of pristine SiO_2 (103.4eV[29]). Since the electronegativity of Si^{4+} (1.80) is higher than that of Ti^{4+} (1.54), the lower binding energy of Si 2p was due to the formation of the Ti-O-Si bonds which caused a less positive charge on Si atoms, as compared to that of pristine SiO_2. This could be an indication of strong interaction of the doped silicon ions with TiO_2 lattice. XPS spectra of films on 3102 Al alloy shown in Fig.S1 also indicates the formation of Ti-O-Si bonds in the TiO_2 films.

The higher photocatalytic performance of these two films might be caused by Si doping. The Ti-O-Si bonds were favorable for transferring photoelectrons from the inner part of particles to the surface and therefore reduced the photoelectron-hole pair recombination possibility[30]. Our previous study shows that TiO_2-SiO_2 nanotubes which were prepared by immersing AAO templates into mixed solution of $(NH_4)_2SiF_6$ and $(NH_4)_2TiF_6$ had higher OH^- concentration than pure TiO_2 nanotubes[31]. In heterogeneous photo-catalysis, the illumination of semiconductor produced electrons (e^-) and holes (h^+). The holes (h^+) were combined with OH^- ions and formed hydroxyl radicals ($h^+ + OH^- \rightarrow \bullet OH$). These surface hydroxyl radicals formed on the surface of the photo-catalyst are oxidizing species which ultimately affected the photocatalytic activity. Therefore, the TiO_2 nanotube films on alumimum alloy had higher photocatalytic oxidation capability.

CONCLUSIONS

(1) AAO templates prepared on the 6061 aluminum alloy substrates had bamboo-like structures with holes, whereas that prepared on pure aluminum and 3102 aluminum alloy substrate had one-dimensional columnar structures.

(2) TiO_2 nanoarray films were prepared in-situ using anodic alumina template by LPD method. The formation of TiO_2 nanotube or nanorod was depended on the volume fraction of Al_2O_3 to AAO template. TiO_2 nanotubes could be formed when the fraction was smaller than 0.71. Otherwise, TiO_2 would exist as nanorods.
(3) TiO_2 nanotube film on aluminum alloy foils exhibited superior photocatalytic activity than that on pure aluminum substrate due to the formation of Ti-O-Si bonds.

ACKNOWLEDGMENTS

This work is financially supported by the National Natural Science Foundation of China (51471006, 51225402), the Beijing Natural Science Foundation (2151001) and the Guangxi Natural Science Foundation (2014GXNSFBB118001).

REFERENCES

1. A. Fujishima, K. Honda,"Electrochemical photolysis of water at a semiconductor electrode," Nature, vol. 238, pp. 37-38, 1972.
2. J.S.Wang, H. Li, H.Y. Li, C. Zhou, H. Wang,"Mesoporous TiO_2 Thin Films Exhibiting Enhanced Thermal Stability and Controllable Pore Size: Preparation and Photocatalyzed Destruction of Cationic Dyes," ACS Appl. Mater. Interfaces, vol. 6, pp. 1623-1631, 2014.
3. N.M. Mahmoodi, M. Arami,"Degradation and toxicity reduction of textile wastewater using immobilized titania nanophotocatalysis," J. Photochem. Photobiol. B., vol. 94, pp. 20-24, 2009.
4. J.S. Wang, H. Li, H.Y. Li, H. Wang,"Thermal Stability and Optimal Photoinduced Hydrophilicity of Mesoporous TiO_2 Thin Films," J. Phys. Chem. C., vol. 116, pp. 9517-9525, 2012.
5. M.N. Ghazzala, H. Kebailib, M. Josepha, D.P. Debeckera, P. Eloya, J.D. Coninckb, E.M. Gaigneauxa,"Photocatalytic degradation of Rhodamine 6G on mesoporous titania films: Combined effect of texture and dye aggregation forms," Appl. Catal. B: Environ., vol. 115– 116, pp. 276-284, 2012.
6. J.S. Wang, H. Li, H.Y. Li, S. Yin, T. Sato,"Preparation and characterization of titania/tetratitanate nanocomposites," Solid State Sci., vol. 11, pp. 988-993, 2009.
7. G.G. Bessegato, J.C. Cardoso, M.M. Zanali,"Enhanced photoelectrocatalytic degradation of an acid dye with boron-doped TiO_2 nanotube anodes," Catal. Today, vol. 240, pp. 100-106, 2015.
8. M.H. Seo, M. Yuasa, T. Kida, J.S. Huh, K. Shimanoe, N. Yamazoe,"Gas sensing characteristics and porosity control of nanostructured films composed of TiO_2 nanotubes," Sens. Actuators, B., vol. 137, pp. 513-520, 2009.
9. R. Jaiswal, J. Bharambe, N. Patel, Alpa, Dashora, D.C. Kothari, A. Miotello, "Copper and Nitrogen co-doped TiO_2 photocatalyst with enhanced optical absorption and catalytic activity," Appl. Catal. B: Environ., vol. 168, pp. 333-341, 2015.
10. C. Colmenares, R. Luque, J.M. Campelo, F. Colmenares, Z. Karpiński, A.A. Romero,"Nanostructured Photocatalysts and Their Applications in the Photocatalytic Transformation of Lignocellulosic Biomass: An Overview," Mater., vol. 2, pp. 2228-2258, 2009.
11. Y.P. Guo, N.H. Lee, H.J. Oh, C.R. Yoon, K.S. Park, W.H. Lee, Y.Z. Li, H.G. Lee, K.S. Lee, S.J. Kim,"Preparation of titanate nanotube thin film using hydrothermal method," Thin Solid Films., vol. 516, pp. 8363-8371, 2008.

12. X.F. Qu, L.X. Cao, F.L. Du,"Fabrication of ordered arrays of CNT/TiO_2 nanotubes and their photocatalytic properties," RSC Adv., vol. 5, pp. 20976-20980, 2015.
13. H.S. Chen, P.H. Chen, S.H. Huang,"Toward highly efficient photocatalysis: a flow-through Pt@TiO_2@AAO membrane nanoreactor prepared by atomic layer deposition," Chem. Comm., vol. 33, pp. 4379-4382, 2014.
14. F. Wang, H.B. Huang, S.G. Yang,"Synthesis of ceramic nanotubes using AAO templates," J. Eur. Ceram. Soc., vol. 29, pp. 1387-1391, 2009.
15. S. Yamanaka, T. Hamaguchi, H. Muta, K. Kurosaki, M. Uno,"Fabrication of oxide nanohole arrays by a liquid phase deposition method," J. Alloys Compd., vol. 373, pp. 312-315, 2004.
16. T. Hamaguchi, M. Uno, K. Kurosaki, S. Yamanaka, "Study on the formation process of titania nanohole arrays," J. Alloys Compd., vol. 386, pp. 265-269, 2005.
17. J. S. Wang, Z. Z. Wang, H. Li, Y. T. Cui, Y. C. Du, "Visible light-driven nitrogen doped TiO_2 nanoarray films: Preparation and photocatalytic activity," J. Alloys Compd., vol. 494, pp. 372-377, 2010.
18. L. H. Ai, C. Y. Zhang, F. Liao, Y. Wang, M. Li, L. Y. Meng and J. Jiang, "Effects of biodiesel, engine load and diesel particulate filter on nonvolatile particle number size distributions in heavy-duty diesel engine exhaust," *J. Hazard. Mater.*, vol.199-200, pp. 282-289, 2012.
19. B. X. Liu, J. S. Wang, J. S. Wu, H. Y. Li, Z. F. Li, M. L. Zhou, T. Y. Zuo, "Controlled fabrication of hierarchical WO_3 hydrates with excellent adsorption performance," J. Mater. Chem. A, vol. 2, pp. 1947-1954, 2014.
20. L. Wang, X. L.Wu, W. H. Xu, X. J. Huang, J. H. Liu, A. W. Xu, "Stable Organic–inorganic hybrid of polyaniline/α-zirconium phosphate for efficient removal of organic pollutants in water environment," ACS Appl. Mater. Interfaces., vol. 4, pp. 2686-2692, 2012.
21. S. Ono, M. Satio, H. Asho," Self-ordering of anodic porous alumina formed in organic acid electrolytes," Electrochim. Acta., vol. 51, pp. 827-833, 2005
22. M. Lu, D. Qin, and H. Li, "Preparation and characterization of porous alumina membrane", J. Lanzhou Univ. Vol. 38, pp. 47- 53,2002.
23. G.E. Thompson, H. Habazaki, K. Shimizu, M. Sakairi, P. Skeldon, X. Zhou, G.C. Wood, "Anodizing of aluminum alloys," Aircraft Eng. & Aerospace Tech., vol. 71, pp. 228-238, 1999.
24. Z. Wang, Y. Zhou, Physical Chemistry, Higher Education Press, Beijing (2004)
25. Y. Masuda, N. Saito, R. Hoffmann, M. R. De Guire, K. Koumoto, "Nano/micro-patterning of anatase TiO_2 thin film from an aqueous solution by site-selective elimination method," Sci. & Tech. Adv. Mater., vol. 4, pp. 461-467, 2003.
26. Z. F. Zhu, Anodic oxidation of aluminium alloy and surface treatment technology, Chemical industry press, Beijing (2004).
27. L. Gao, S. Zheng, Q. Zhang, TiO_2 nanophtocatalyst and its application, Chemical industry press, Beijing (2002).
28. X. L. Yan, J. He, D.G. Evans, X. Duan, Y. Zhu, " Preparation, characterization and photocatalytic activity of Si-doped and rare earth-doped TiO_2 from mesoporous precursors," Appl. Catal.B: Environ., vol. 55, pp. 243-252, 2005.
29. C. Hu, Y. Z. Wang, H. X. Tang. "Preparation and characterization of surface bond-conjugated TiO_2/SiO_2 and photocatalysis for azo dyes," Appl. Catal. B: Environ., vol. 30, pp. 277-285, 2001.

30. M. Fujishima, H Takatori, H Tada. "Interfacial chemical bonding effect on the photocatalytic activity of TiO_2–SiO_2 nanocoupling systems," J. Colloid & Interf. Sci., vol. 361, pp. 628-631, 2011.
31. Z. Z Wang, J. S. Wang, H. Y. Li, G. S. Sun, and K. L. Huang, "Fabrication and photocatalytic activity of TiO2/SiO2 composite nanotubes," Res. Chem. Intermediat. vol. 37, pp. 541-549, 2011.

Table 1. The correlation of morphology of TiO_2 nanoarrays with volume ratio of alumina in different AAO membranes

Electrolyte	H_2SO_4	H_2SO_4, H_3PO_4	$H_2C_2O_4$	$H_2C_2O_4$, H_3PO_4	H_3PO_4,
Volume ratio	0.81	0.23	0.92	0.70	0.56
Morphology of TiO_2	Nanorod	Nanotube	Nanorod	Nanotube	Nanotube

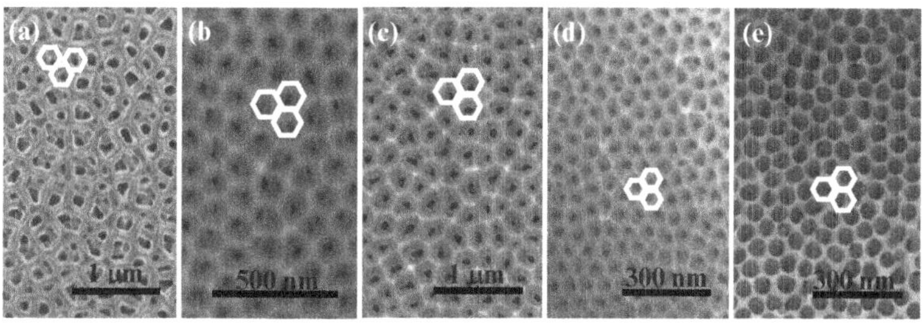

Fig. 1. SEM images of the AAO templates anodized with different electrolytes. (a) H_3PO_4, (b) $H_2C_2O_4$, (c) Mixed electrolytes of $H_2C_2O_4$ and H_3PO_4, (d) H_2SO_4 and (e) Mixed electrolytes of H_2SO_4 and H_3PO_4.

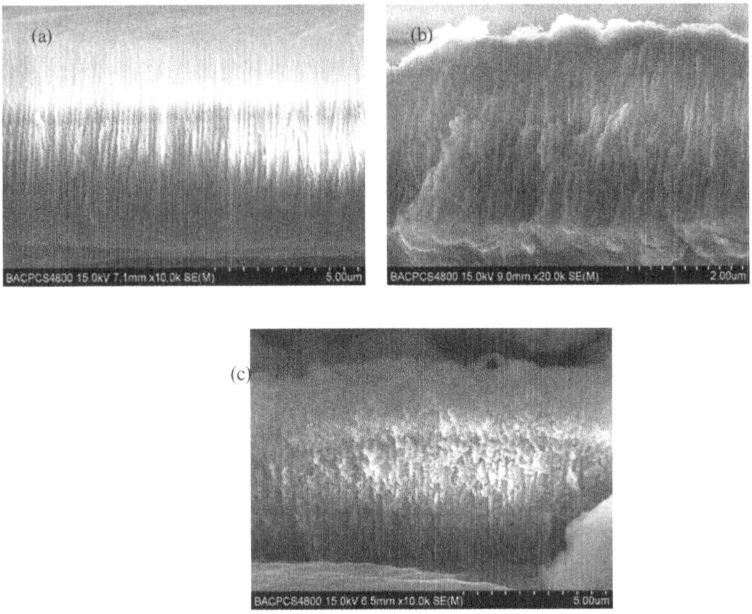

Fig.2 SEM morphology of AAO templates. a) pure aluminum, b) 3102 aluminum alloy and c) 6061 aluminum alloy.

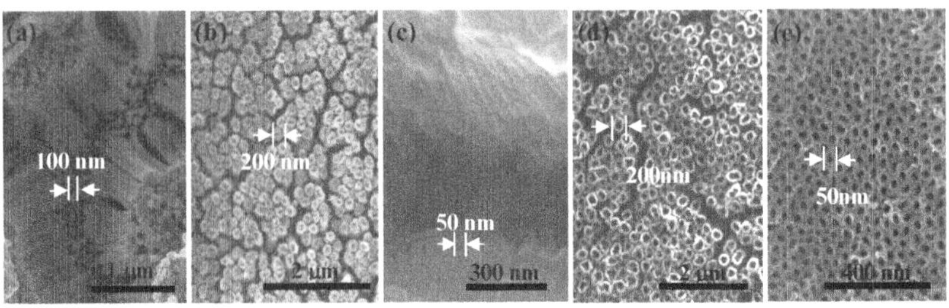

Fig. 3. SEM images of the TiO_2 nanoarrays prepared with AAO templates which were anodized with different electrolytes. (a) $H_2C_2O_4$, (b) Mixed electrolytes of $H_2C_2O_4$ and H_3PO_4, (c) H_2SO_4, (d) H_3PO_4 and (e) Mixed electrolytes of H_2SO_4 and H_3PO_4.

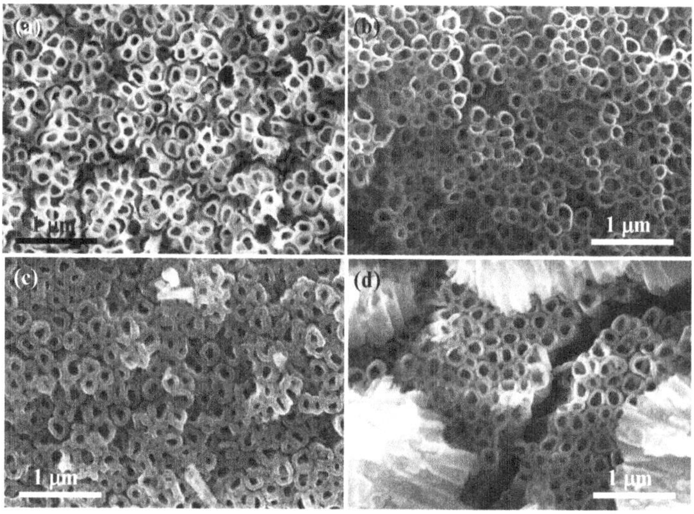

Fig.4. SEM images of TiO_2 nanotubes prepared at different LPD temperature. (a) 30 °C, (b) 35 °C, (c) 40 °C and (d) 45 °C.

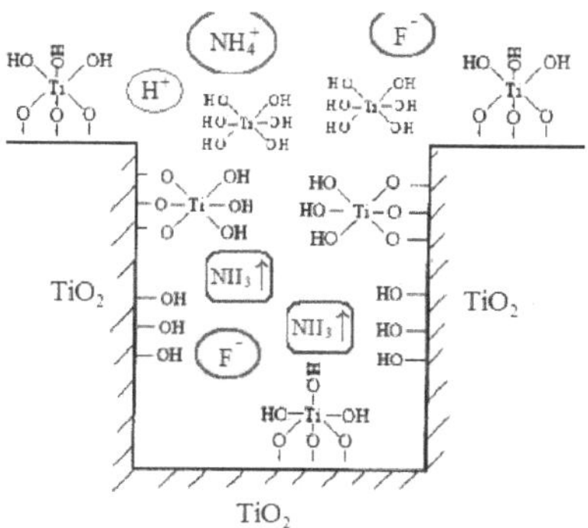

Fig. 5. The schematic diagram of LPD process at high temperature.

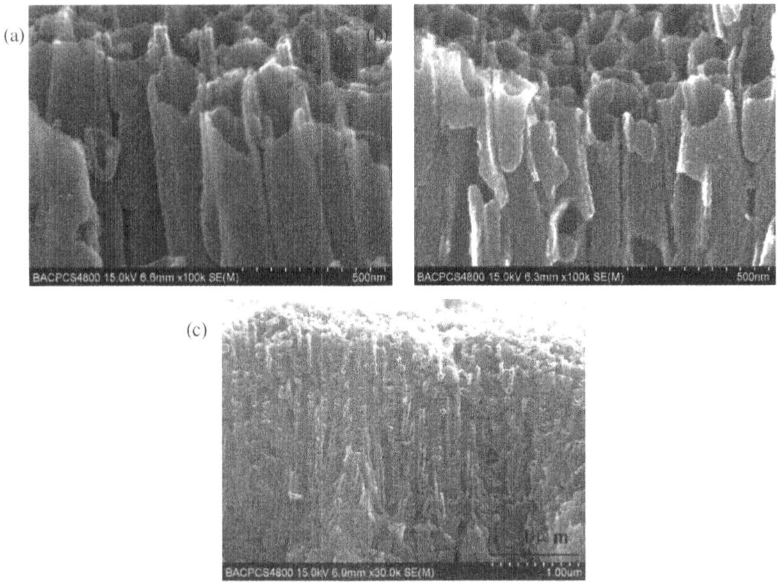

Fig.6 SEM morphology of TiO_2 films on different substrates. (a) pure aluminum, (b)3032 aluminum alloy and (c) 6061 aluminum alloy.

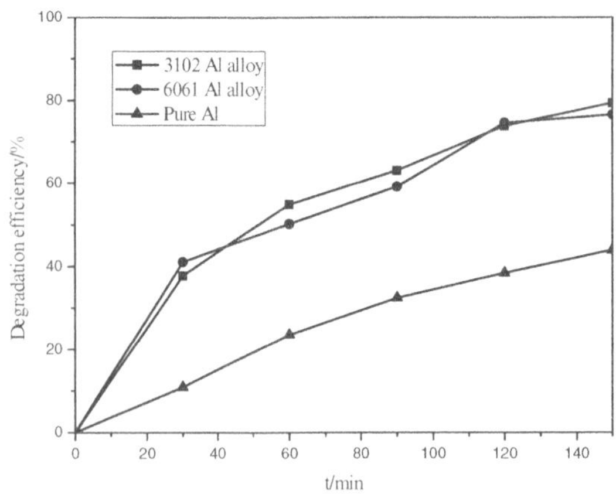

Fig.7 Photocatalytic property of TiO_2 thin films fabricated on aluminum alloy and pure aluminum substrates.

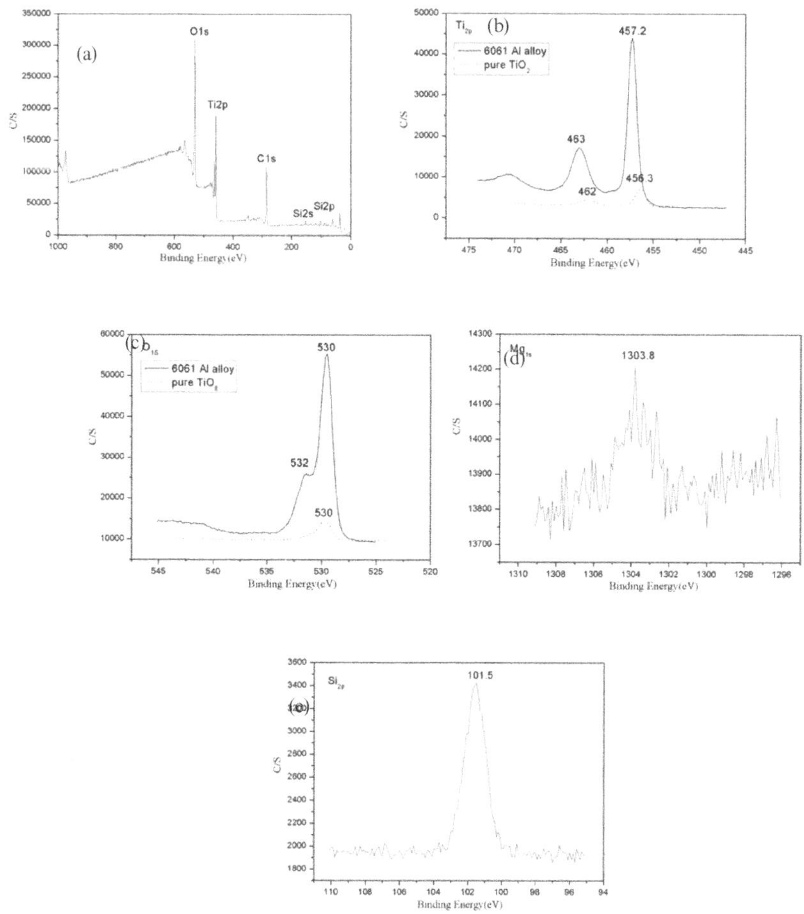

Fig.8 XPS spectra of TiO_2 nanofilms on 6061 Al alloy. a)full spectra, b)Ti_{2p}, c) O_{1s}, d) Mg_{1s} and e) Si_{2p}.

Advanced Functional Materials, Devices, and Systems for the Environment

ELECTROCHEMICAL DEVICES WITH OXIDE ION ELECTROLYTES FOR FORMATION OF HYDROGEN AND DECOMPOSITION OF CARBON DIOXIDE FROM THE CH_4–CO_2 MIXED BIOGAS

Yoshihiro Hirata, Soichiro Sameshima, Taro Shimonosono
Department of Chemistry, Biotechnology, and Chemical Engineering, Kagoshima University, 1-21-40 Korimoto, Kagoshima 890-0065, Japan

ABSTRACT

This paper reports the usefulness of porous electrochemical devices with oxide ion electrolytes for the production of a H_2 fuel by (1) reforming of biogas with a typical composition of 60% CH_4–40% CO_2 ($CH_4 + CO_2 \rightarrow 2CO + 2H_2$) and by (2) the shift reaction of CO with H_2O ($CO + H_2O \rightarrow H_2 + CO_2$). Reforming of CH_4 with CO_2 proceeded through a Ni–Gd-doped ceria (GDC) cathode / GDC porous electrolyte / Ru–GDC anode cell operated at 400–800°C. The supplied CO_2 was changed to CO and O^{2-} ions by the reaction with electrons at the cathode ($CO_2 + 2e^- \rightarrow CO + O^{2-}$). The produced CO and O^{2-} ions were transported to the anode through a porous GDC film. In the anode CH_4 reacted with O^{2-} ions to produce CO, H_2 and electrons ($CH_4 + O^{2-} \rightarrow CO + H_2 + 2e^-$). A similar electrochemical reaction occurred at 400–500°C in the CO–H_2O system. The CO gas produced from a biogas was decomposed to solid carbon and O_2 gas using an electrochemical cell of the Ni–8 mol% yttria-stabilized zirconia (YSZ) cathode / dense YSZ electrolyte / Ru–YSZ anode system at 800°C. The overall reaction produces a clean H_2 fuel, oxygen gas and solid carbon, and no CO_2 is exhausted.

INTRODUCTION

The concentration of CO_2 in atmosphere increases since the end of 18th century because of the consumption of a large amount of coal, oil and natural gas and this phenomenon causes a serious problem of global warming. To overcome this situation, renewable energies such as wind power, solar energy, water power, geothermal energy, tidal power, and biomass energy have been intensively developed as alternative energies. A large amount of biogas can be produced from waste food or drainage in Japan. This gas contains 50–70 vol% CH_4, 30–50 vol% CO_2 and a small amount of H_2S and NH_3 gases. The reforming of CH_4 with CO_2 by Eq. (1) is an attractive process to produce H_2 and CO fuels for solid oxide fuel cell (SOFC) or polymer electrolyte fuel cell (PEFC)[1, 2]: $CH_4 + CO_2 \rightarrow 2H_2 + 2CO$ (1). On the other hand, the following photosynthesis by plants produces grape sugar, H_2O and O_2 gas from CO_2 and H_2O: $12H_2O + 6CO_2$ + photoenergy $\rightarrow C_6H_{12}O_6$ (grape sugar) $+ 6H_2O + 6O_2$ (2). This photosynthesis process is ideal to reduce the amount of CO_2 gas. However, no commercial system of artificial photosynthesis is established at this moment. Recently, our group developed the electrochemical cells of Ru–Gd-doped ceria (GDC) anode/porous GDC electrolyte/Ni or Ru–GDC cathode system[3–7] to reform biogas (mixed gas of 30–70 vol% CH_4 and 70–30vol% CO_2). These cells work well under a small electric field strength of 1–2 V/cm at 400–800°C. Furthermore, the similar types of electrochemical cells work well to proceed the shift reaction of CO gas with H_2O vapor ($CO + H_2O \rightarrow H_2 + CO_2$)[6, 8, 9] and decomposition of CO gas into solid carbon and O_2 gas ($2CO \rightarrow 2C + O_2$)[10, 11]. This paper reviews the main results of reforming of CH_4–CO_2 mixed gas, shift reaction of CO–H_2O system and decomposition of CO gas using developed electrochemical cells. The combined reaction system can produce a H_2 fuel with no formation of CO_2 gas from a biogas and is explained in Section "COMBINED REACTION SYSTEM USING ELECTROCHEMICAL CELLS": Biogas ($3CH_4$ +

$2CO_2) + CO_2$ (circulated) $+ H_2O \rightarrow 7H_2 + CO_2$ (circulated) $+ 5C + 5/2O_2$ (3).

REFORMING OF CH_4–CO_2 MIXED GAS AND SHIFT REACTION OF CO WITH H_2O VAPOR USING ELECTROCHEMICAL CELLS

The biogas produced by CH_4 fermentation of waste food or drainage contains 50–70% CH_4 and 30–50% CO_2[6]. Figure 1(a) and (b) show the possible reaction mechanisms for the reforming of CH_4–CO_2 mixed gas and the shift reaction of CO–H_2O mixed gas through porous GDC cells, respectively. Ni and Ru in the cathode in Fig. 1(a) worked well to reduce CO_2 to CO plus O^{2-} ions by the reaction with supplied electrons ($CO_2 + 2e^- \rightarrow CO + O^{2-}$). The produced CO and O^{2-} ions are transported to the anode through a porous GDC layer. In the anode with Ru or Cu catalyst, CH_4 reacts with O^{2-} ions to produce H_2, CO and electrons ($CH_4 + O^{2-} \rightarrow 2H_2 + CO + 2e^-$). The detailed results were already reported in references[3–7]. Further shift reaction of CO with H_2O is presented in Fig. 1(b). H_2O gas in the cathode reacts with the supplied electrons to form H_2 and O^{2-} ions through the catalytic reactions of metal by Eqs. (12) and (13) in Fig. 1. The formed O^{2-} ions and CO gas are transported to the anode through a porous GDC layer. In the anode, CO gas reacts with O^{2-} ions to produce CO_2 and electrons through catalytic reactions by Eqs. (15) and (16). The combination of Eqs. (11) and (14) yields the shift reaction by Eq. (17). The CO_2 gas separated from a H_2 fuel is mixed again with CH_4-rich biogas. The coupling of reforming of biogas (Eq. (10)) and shift reaction of CO with H_2O (Eq. (17)) produces the H_2–CO_2 mixed gas by Eq. (18) in Fig. 1.

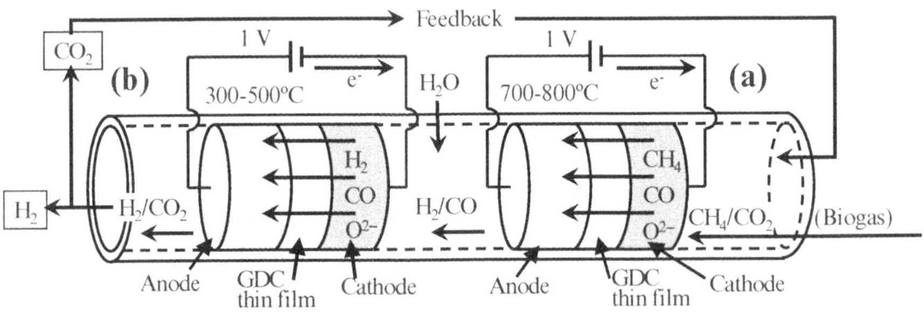

GDC : $Ce_{0.8}Gd_{0.2}O_{1.9}$ (Gd-doped ceria electrolyte, 50 μm in thickness)

$CO + O^{2-} \rightarrow CO_2 + 2e^-$ (14) (CO oxidation)

$H_2O + 2e^- \rightarrow H_2 + O^{2-}$ (11) (H_2O reduction)

$CH_4 + O^{2-} \rightarrow CO + 2H_2 + 2e^-$ (7) (CH_4 oxidation)

$CO_2 + 2e^- \rightarrow CO + O^{2-}$ (4) (CO_2 reduction)

$M + O^{2-} \rightarrow MO + 2e^-$ (15), $CO + MO \rightarrow M + CO_2$ (16)

$M + H_2O \rightarrow MO + H_2$ (12), $MO + 2e^- \rightarrow M + O^{2-}$ (13)

$M + O^{2-} \rightarrow MO + 2e^-$ (8), $CH_4 + MO \rightarrow CO + 2H_2 + M$ (9)

$M + CO_2 \rightarrow MO + CO$ (5), $MO + 2e^- \rightarrow M + O^{2-}$ (6)

(11) + (14) $CO + H_2O \rightarrow CO_2 + H_2$ (17)

(4) + (7) $CH_4 + CO_2 \rightarrow 2H_2 + 2CO$ (10)

Total reaction (10) + (17) $CH_4 + CO_2 + 2H_2O \rightarrow 4H_2 + 2CO_2$ (18)

Fig. 1 Possible reaction mechanism for (a) reforming of CH_4–CO_2 mixed gas and (b) shift reaction of CO–H_2O mixed gas through porous GDC cells.

Figure 2 shows the fractions of outlet gas through the porous Gd-doped ceria electrolyte (GDC, $Ce_{0.8}Gd_{0.2}O_{1.9}$) with Co–GDC cathode and Fe–GDC anode at 400–700°C where the inlet gas of CO/H_2O = 1/7.35 volume ratio was passed[8]. This cell produced a large amount of H_2 and CO_2 gases with the H_2/CO_2 volume ratio ≈ 1. The fraction of H_2 fuel was relatively high at a lower heating temperature, which agreed with the prediction from the thermodynamic calculation[8]. After the measurement of gas fraction at a constant temperature, the applied voltage was switched off and the supplied gases were substituted for an Ar gas. After the heating temperature reached the next reaction temperature, CO gas and H_2O vapor were supplied again and the applied voltage was switched on. Therefore, the gas composition gradually approached the steady state with reaction time. After the shift reaction at 700°C, as-produced Co_3O_4 in the GDC matrix was reduced to Co metal, which acted as a catalyst to form a H_2 fuel. After the shift reaction, four phases of Fe, Fe_3O_4, $GdFeO_3$ and GDC were identified in the anode.

Based on the above results, Eqs. (19)–(21) represent the possible reactions at the cathode.

$$Co + H_2O \rightarrow CoO + H_2 \quad (19)$$
$$CoO + 2e^- \rightarrow Co + O^{2-} \quad (20)$$
$$(19) + (20) \quad H_2O + 2e^- \rightarrow H_2 + O^{2-} \quad (21)$$

Co works as a catalyst to reduce H_2O. The formed H_2 fuel, O^{2-} ions and supplied CO gas are transported to the anode through the porous GDC layer. In the Fe–GDC anode, the following reactions explain the phases produced in the anode.

$$3Fe + 4O^{2-} \rightarrow Fe_3O_4 + 8e^- \quad (22)$$
$$Fe_3O_4 + CO \rightarrow 3FeO + CO_2 \quad (23)$$
$$3FeO + O^{2-} \rightarrow Fe_3O_4 + 2e^- \quad (24)$$
$$(23) + (24) \quad CO + O^{2-} \rightarrow CO_2 + 2e^- \quad (25)$$

The change of Fe valence ($Fe^{2+} \leftrightarrow 2Fe^{3+}$) in Fe_3O_4 accelerates the reaction between CO gas and O^{2-} ions through Eqs. (23) and (24).

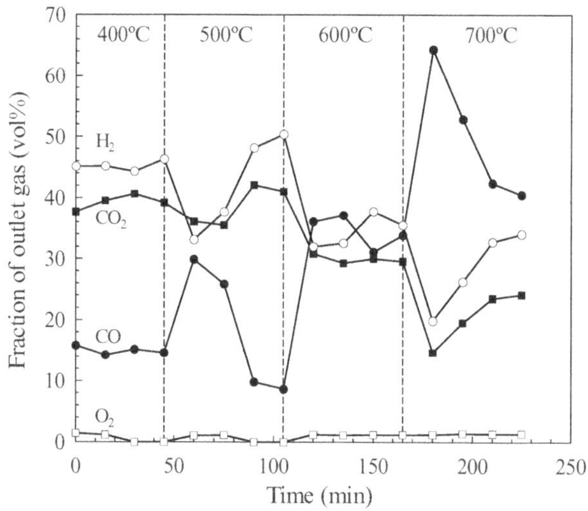

Fig. 2 Fractions of CO, H_2, CO_2 and O_2 gases through a porous electrochemical cell at 400–700°C[8].

ELECTROCHEMICAL DECOMPOSITION OF CO AND CO_2 GASES

Our next challenge is to decompose CO or CO_2 gas formed in dry reforming of CH_4–CO_2

mixed gas or shift reaction into solid carbon and oxygen gas. In our previous study[10], CO and CO_2 gases were electrochemically decomposed to solid carbon and O_2 gas using a porous cell of Ni–YSZ cathode/porous YSZ electrolyte/Ru–YSZ anode. In such a cell with a porous electrolyte (Fig. 3), a reactant gas was supplied to the cathode side and a formed gas was exhausted from the anode side through the porous electrolyte. The decomposition fractions of CO and CO_2 gases increased at a higher temperature (400–800°C) and at a lower gas flow rate. In the case of CO_2 decomposition, CO_2 and O_2 gases were identified in the outlet gas. The supplied CO_2 gas was decomposed into oxide ions and CO gas ($CO_2 + 2e^- \rightarrow CO + O^{2-}$, $CO_2 + 4e^- \rightarrow C + 2O^{2-}$) or solid carbon in the cathode, and the formed CO gas reacted with the formed oxide ions to return to CO_2 gas in the anode ($CO + O^{2-} \rightarrow CO_2 + 2e^-$, $2O^{2-} \rightarrow O_2 + 4e^-$). Similarly, CO, O_2 and CO_2 gases were identified in the outlet gas of CO decomposition. Further improvement is the increase of decomposition fraction of CO or CO_2 gas and the separation of formed O_2 gas from the supplied CO or CO_2 gas. When a dense oxide ion electrolyte was used for the decomposition of CO or CO_2 gas, the formed O_2 gas was completely separated from the supplied gas and solid carbon formed in the cathode (Fig. 3)[11].

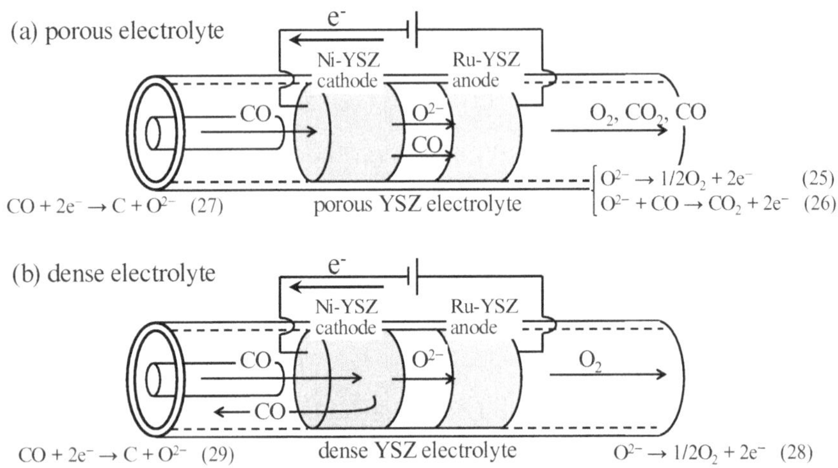

Fig. 3 Scheme of electrochemical decomposition of CO and CO_2 gases using (a) porous oxide ion electrolyte and (b) dense oxide ion electrolyte.

Figure 4 shows (a) the O_2 content of outlet gas (carrier gas: Ar gas) and (b) the decomposition rate of CO gas (10 vol% CO–90 vol% Ar mixed gas at 5 ml/min) at 800°C as a function of applied voltage using a cell of dense 8 mol% yttria-stabilized zirconia electrolyte (40 μm thickness)[11]. The decomposition rate of CO was determined from the amount of O_2 formed in the anode ($2CO \rightarrow 2C + O_2$). The O_2 gas content of 5.4% of the outlet gas corresponded to the decomposition rate of 100% of the supplied CO gas. Only O_2 gas was detected in the anode, indicating the complete separation of O_2 gas from the supplied CO gas. The decomposition rate of CO was in the range of 9–15% at an applied voltage of 1.0 V, but drastically increased to 67–100% at 2–8 V. Figure 5 shows the X-ray diffraction patterns of Ni–YSZ cathode and Ru–YSZ anode after the CO decomposition at 800°C under an applied voltage of 8.0 V. No change of Ni and YSZ electrolyte was observed in the cathode during the CO decomposition. In the anode, a part of Ru was oxidized to RuO_2 by the reaction with O^{2-} ions transported through the dense YSZ film. The elementary distribution in the cathode near the dense YSZ electrolyte after the CO decomposition was analyzed by electron probe microanalyzer. The carbon deposition occurred along the interface between the distribution maps of Ni element and Zr element[11]. The concentration of deposited carbon was 13.2 mass% near the electrolyte/cathode interface, 2.6 mass% at the center of the cathode, and 1.2 mass% near the cathode surface, respectively. That is, the carbon deposition in the cathode became difficult to occur with increasing distance from the electrolyte interface. In the cathode, Ni metal reacts with CO to form NiO and carbon (Eq. (30)). The formed NiO is decomposed to Ni and O^{2-} ions by reaction with the supplied electrons (Eq. (31)). The produced O^{2-} ions are transported to the anode through the dense YSZ electrolyte. The summation of Eqs. (30) and (31) provides Eq. (32) (i.e. the electrochemical reduction of CO gas).

$$Ni + CO \rightarrow NiO + C \quad (30)$$
$$NiO + 2e^- \rightarrow Ni + O^{2-} \quad (31)$$
$$(30) + (31) \quad CO + 2e^- \rightarrow C + O^{2-} \quad (32)$$

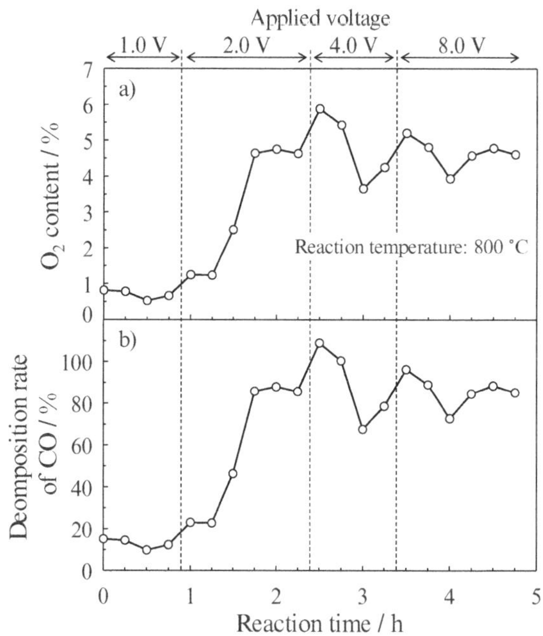

Fig. 4 (a) O_2 content of outlet gas in anode side and (b) decomposition rate of CO as a function of applied voltage[11].

In the anode, a part of the transported O^{2-} ions releases the electrons to form O_2 gas (Eq. (33) with a reaction rate of x). The other O^{2-} ions react with Ru metal to form RuO_2 and electrons (Eq. (34) with a reaction rate of (1−x)). The summation of Eqs. (33) and (34) provides Eq. (35) (i.e. the formation of O_2 gas and RuO_2 due to the electrochemical oxidation by O^{2-} ions).

$$2xO^{2-} \rightarrow xO_2 + 4xe^- \quad (33)$$
$$(1-x)Ru + 2(1-x)O^{2-} \rightarrow (1-x)RuO_2 + 4(1-x)e^- \quad (34)$$
$$(33) + (34) \quad 2O^{2-} + (1-x)Ru \rightarrow xO_2 + (1-x)RuO_2 + 4e^- \quad (35)$$

The total reaction for the CO decomposition in the electrochemical cell is expressed by Eq. (36).

$$2CO + (1-x)Ru \rightarrow 2C + xO_2 + (1-x)RuO_2 \quad (36)$$

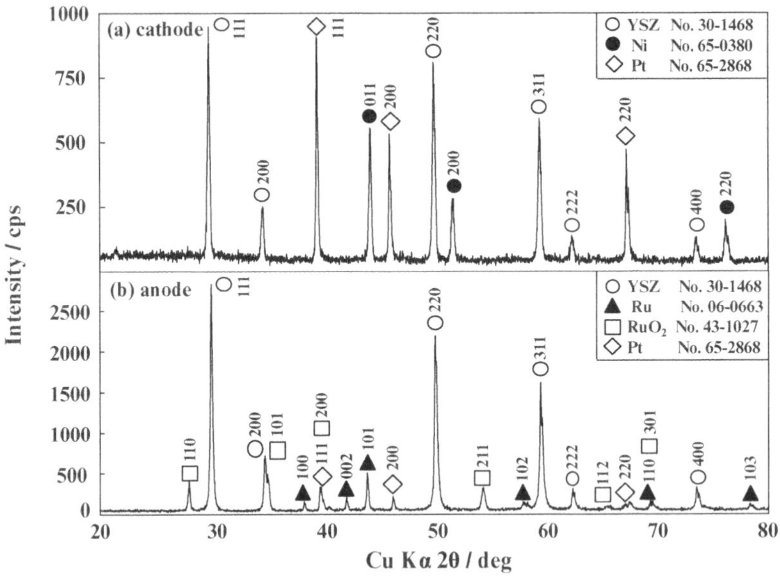

Fig. 5 X-ray diffraction patterns of (a) Ni–YSZ cathode and (b) Ru–YSZ anode after the CO decomposition experiment at 800°C.

Figure 6 shows the outlet gas compositions of (a) cathode side and (b) anode side during CO_2 decomposition at 800°C as a function of applied voltage[11]. The outlet gas in the cathode side contained about 93% CO_2 and 7% CO gas, and the composition was not sensitive to the applied voltage. No change of Ni and YSZ was observed in the cathode. A part of Ru metal in the anode was oxidized to RuO_2. A small amount of carbon was deposited in the cathode. The carbon concentration was 2.5 mass% near the cathode surface, 1.6 mass% at the center of cathode and 2.1 mass% near the electrolyte interface, respectively. The possible decomposition mechanisms of CO_2 gas are proposed by Eqs. (37) and (38).

$$CO_2 \xrightarrow{y} CO + 1/2O_2 \quad (37)$$

$$CO_2 \xrightarrow{q} C + O_2 \qquad (38)$$

When the reaction rates of Eqs. (37) and (38) are y and q, respectively, the molar fractions of unreacted CO_2 gas, produced CO and O_2 gases are expressed to be (1−y−q), y and (y/2+q), respectively. The reaction rates (y and q) were calculated using the measured gas compositions of CO_2, CO and O_2. Figure 7 shows the variations of y and q values as a function of applied voltage. The y value was in the range of 6.5–7.3% and almost independent of the applied voltage, while the q value was almost 0% at 1.0 V and increased to about 3% at 2.0–8.0 V. Thus, CO formation occurs predominantly in the wide applied voltage range. On the other hand, an applied voltage larger than 1.0 V is needed for the carbon deposition. The increase of the applied voltage during the electrodecomposition of CO_2 influences the decomposition mechanisms (Fig. 7).

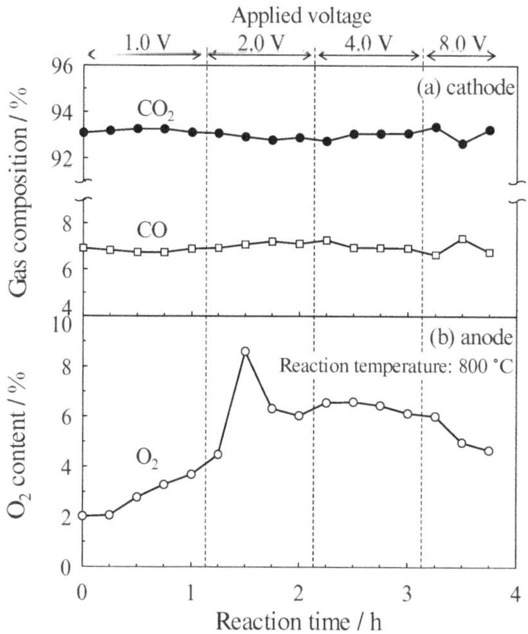

Fig. 6 Outlet gas composition of (a) cathode side and (b) anode side during CO_2 decomposition at 800°C[11].

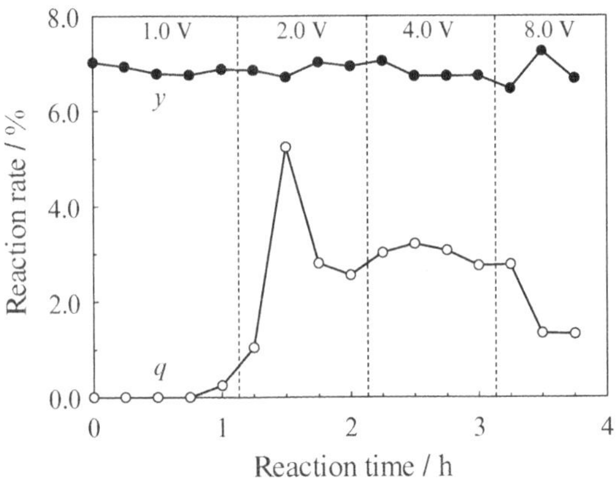

Fig. 7 Variations of y (conversion rate of CO_2 to CO) and q (conversion rate of CO_2 to C) values in Eqs. (37) and (38) as a function of applied voltage[11].

COMBINED REACTION SYSTEM USING ELECTROCHEMICAL CELLS

It was found that the developed electrochemical cells accelerate the following basic reactions.

Dry reforming: $CH_4 + CO_2 \rightarrow 2H_2 + 2CO$ (39)
Shift reaction: $CO + H_2O \rightarrow H_2 + CO_2$ (40)
CO decomposition: $2CO \rightarrow 2C + O_2$ (41)

The combination of above three equations for a biogas with a typical composition of 60% CH_4–40% CO_2 provides Eq. (42).

$$\text{Biogas } (3CH_4 + 2CO_2) + CO_2 \text{ (circulated)} + H_2O \rightarrow 7H_2 + CO_2 \text{ (circulated)} + 5C + 5/2O_2 \quad (42)$$

This combined equation produces a H_2 fuel with no formation of CO_2 gas. This combined process is also effective for 100% CH_4 fuel as expressed by Eq. (43).

$$CH_4 + H_2O + CO_2 \text{ (circulated)} \rightarrow 3H_2 + C + 1/2O_2 + CO_2 \text{ (circulated)} \quad (43)$$

This combined reaction system is to be studied using a natural biogas. Also in our previous study[12], the reformed mixed fuel of H_2 and CO gases were supplied to an SOFC composed of a dense GDC electrolyte ($Ce_{0.8}Gd_{0.2}O_{1.9}$). The open-circuit voltage and a maximum power density at 800°C were comparable for both the reformed fuel and for a pure H_2 fuel. As described above, the electrochemical devices with oxide ion electrolytes can be applied for a biogas to produce a clean H_2 fuel or a H_2–CO mixed fuel and the remaining CO gas can be decomposed into solid carbon and O_2 gases.

CONCLUSIONS

Electrochemical reforming of CH_4 with CO_2 in the volume ratio of CH_4/CO_2 = 1 proceeded through in a Ni–GDC cathode/porous GDC electrolyte/Ru–GDC anode cell to form a

H_2–CO mixed fuel. A Co–GDC cathode/porous GDC electrolyte/Fe–GDC anode cell produced a H_2-enriched mixed gas of 15% CO–45% H_2–40% CO_2 through the electrochemical shift reaction of CO with H_2O. A Ni–YSZ cathode/dense YSZ electrolyte/Ru–YSZ anode cell at 800°C decomposed CO gas of 70–100% into solid carbon and oxygen at 2.0–8.0 V of applied voltage. In the case of CO_2 decomposition at 800°C, the outlet gas from the cathode contained 93% CO_2 gas and 7% CO gas. The combination of electrochemical cells with oxide ions can produce a clean H_2 fuel with no formation of CO_2 gas.

REFERENCES

[1]Sameshima, S., Hirata, Y., Sato, J., Matsunaga, N., "Synthesis of hydrogen and carbon monoxide from methane and carbon dioxide", J. Ceram. Soc. Japan, vol. 116, pp. 374–379, 2008.

[2]Sameshima, S., Hirata, Y., Hamasaki, K., Ohshige, H., Matsunaga, N., "Synthesis of hydrogen–carbon monoxide fuel from methane–carbon dioxide mixed gases", J. Ceram. Soc. Japan, vol. 117, pp. 630–634, 2009.

[3]Hirata, Y., Terasawa, Y., Matsunaga, N., Sameshima, S., "Development of electrochemical cell with layered composite of the Gd-doped ceria electronic conductor system for generation of H_2–CO fuel through oxidation–reduction of CH_4–CO_2 mixed gases", Ceram. Inter., vol. 35, pp. 2023–2028, 2009.

[4]Matayoshi, S., Hirata, Y., Sameshima, S., Matsunaga, N., Terasawa, Y., "Electrochemical reforming of CH_4–CO_2 gas using porous Gd-doped ceria electrolyte with Ni and Ru electrodes", J. Ceram. Soc. Japan, vol. 117, pp. 1147–1152, 2009.

[5]Ando, M., Hirata, Y., Sameshima, S., Matsunaga, N., "Electrochemical reforming of CH_4–CO_2 mixed gas using porous yttria-stabilized zirconia cell", J. Ceram. Soc. Japan, vol. 119, pp. 794–800, 2011.

[6]Hirata, Y., Matsunaga, N., Sameshima, S., "Reforming of biogas using electrochemical cell", J. Ceram. Soc. Japan, vol. 119, pp. 763–769, 2011.

[7]Suga, Y., Yoshinaga, R., Matsunaga, N., Hirata, Y., Sameshima, S., "Electrochemical reforming of CH_4–CO_2 mixed gas using Gd-doped ceria porous electrolyte with Cu electrode", Ceram. Inter., vol. 38, pp. 6713–6721, 2012.

[8]Hirata, Y., Kisanuki, Y., Sameshima, S., Shimonosono, T., "Formation of hydrogen by electrochemical reaction of CO gas and H_2O vapor using porous Gd-doped ceria electrolyte", Ceram. Inter., vol. 40, pp. 10153–10157, 2014.

[9]Ueda, K., Hirata, Y., Sameshima, S., Shimonosono, T., Yamaji, K., "Formation of hydrogen from the CO–H_2O system using porous Gd-doped ceria electrochemical cell with MnO cathode and Fe_3O_4 anode, J. Asian Ceram. Soc., vol. 3, pp. 82–87, 2015.

[10]Hirata, Y., Ando, M., Matsunaga, N., Sameshima, S., "Electrochemical decomposition of CO_2 and CO gases using porous yttria-stabilized zirconia cell", Ceram. Inter., vol. 38, pp. 6377–6387, 2012.

[11]Shimonosono, T., Ueno, M., Hirata, Y., Sameshima, S., "Decomposition of carbon monoxide and carbon dioxide using cathode-supported dense yttria-stabilized zirconia electrolyte cell", Ceram. Inter., vol. 40, pp. 15957–15964, 2014.

[12]Furukawa, N., Hirata, Y., Sameshima, S., Matsunaga, N., "Evaluation of electric power of SOFC using reformed biogas", Materials Science Forum, vol. 761, pp. 11–14, 2013.

GASTIGHT, CLOSED PORE INCLUSIVE POROUS CERAMICS THROUGH A SUPERPLASTICALLY FOAMING METHOD

Akira Kishimoto*, Atsuki Tohji, Takashi Teranishi and Hidetaka Hayashi

Graduate School of Natural Science and Technology, Okayama University

Division of Chemistry and Biochemistry, Graduate School of Natural Science and Technology, Okayama University, 3-1-1 Tsushima-naka, Kita-ku, Okayama 700-8530, Japan

ABSTRACT

Our innovative superplastically foaming method was used to create closed-pore inclusive zirconia-based ceramics. Yttrium oxide was added to monoclinic zirconia to stabilize the matrix. Sintering and superplastic deformation were used to form a solid solution and transform it to a cubic phase. Porous ceramics formed via this post-stabilization method exhibited anomalous thermal expansion due to the silica derived from the foaming agent SiC. This anomalous thermal expansion disappeared following substitution of the SiC foaming agent for AlN. The resulting superplastically foamed porous ceramics exhibited large thermal conductivities, comparable to that of the fully densified ceramics, but with a porosity as great as 35%. This high thermal conductivity was attributed to the alumina derived from the foaming agent AlN. To suppress the formation of alumina, the samples were pre-sintered in an argon atmosphere. The modified superplastically foamed ceramics exhibited smaller pores, with a porosity of 33%, as well as favorable thermal insulation, with a thermal conductivity that was approximately half that of the fully dense material.

INTRODUCTION

Porous ceramics incorporate pores to improve the material properties, especially the thermal insulation, maintaining the inherent ceramic properties of corrosion resistance and high mechanical strength. We have previously fabricated porous ceramics by introducing pores following full densification of the matrix.[1-19] Using this method, granular foaming agents were arranged at regular intervals in the matrix powder prior to compaction. Following densification of the matrix, the pores expanded utilizing the superplastic deformation driven by the gas pressure evolved from the foaming agents.[20-24] This process is termed superplastically foaming method.

Almost all pores were closed and surrounded by fully densified walls. As a result, these composites are expected to exhibit high mechanical strength, gas-tightness and thermal insulation, which have already been demonstrated. These properties are in contrast with those of conventional porous ceramics fabricated through partial sintering. Porous ceramics containing closed pores are promising as high-temperature gas-tight thermal insulators.

The degree of pore expansion depends on the superplastic deformation limit and the rate whereby the gas evolution remains constant. To increase the porosity of superplastically foamed ceramics, dispersoids such as silica[3-7, 13] and spinel[8,15] may be added to the matrix. These dispersoids do not necessarily have a positive effect on the properties of porous ceramics, however, even if they increase the porosity.

We have previously fabricated suplerplastically foamed silica dispersed 3YSZ (3-mol% yttria-stabilized zirconia) porous ceramics. The thermal conductivity of the superplastically foamed ceramics decreased from 3.5 $Wm^{-1}K^{-1}$ for the fully dense material to 1.7 $Wm^{-1}K^{-1}$ for the foamed ceramic with a porosity of 25%.[6] Although the thermal conductivity reduced to approximately half that of fully dense material, the silica-dispersed 3YSZ exhibited a larger thermal conductivity than the 3YSZ without the dispersoid. This can be ascribed to the presence of silica at the gain boundaries.

We have also reported the fabrication of superplastically foamed zirconia ceramics, which were stabilized using both lanthanum and yttrium.[19] A low thermal conductivity was obtained with a cubic phase; however, this exhibited little superplastic deformation. To overcome this drawback, we developed a post-stabilization method, whereby sintering and superplastic deformation were carried out with monoclinic or tetragonal YSZ dispersed with lanthanum oxide, followed by solid solution to transform to a cubic phase with a lower thermal conductivity.

With 8YSZ, this post-stabilization method could be adopted by dispersing yttrium oxide into monoclinic or tetragonal zircona that was partially stabilized using yttria. In this work, monoclinic zirconia powder dispersed with yttria powder was used to form a powder containing a foaming agent. Expansion of the pores occurred during the sintering/foaming process, which was followed by stabilization, leading to the formation of a cubic phase. We report here the mechanical and thermal properties as well as the thermal insulation properties.

EXPERIMENTAL

Three different powders were used with the post-stabilization superplastically foaming method: matrix monoclinic zirconia powder, 0YSZ (TZ-0, Tosoh Co. Ltd., Tokyo, Japan) and stabilization oxide yttria (GR-F, Shin-Etsu Chemical Co. Ltd., Tokyo, Japan), as well as the foaming agent (β-SiC, ABCR, GmbH & Co., Germany, or AlN, JC, Toyo Aluminium K.K., Osaka, Japan). Fully stabilized zirconia 8YSZ (TZ8, Tosoh, Tokyo, Japan) was used as the dense ceramics material, and graphite pore former was used as the conventional porous ceramics.

Compaction followed by sintering was used to form superplastically foamed porous ceramics containing only closed pores. We have previously fabricated macroscopic 8YSZ monofoam pellets (20 mm in diameter), which contained only one pore,[5] as well as superplastically foamed porous 8YSZ,[6] using dispersed silica as the agent to facilitate superplasticity. 8YSZ powder containing 10-mol% silica was chosen to provide a balance of superplastic deformation ability and the thermal expansion profile.

The porosity was controlled by adjusting the SiC/matrix ratio. The powder mixtures were pressed uniaxially using a 20-mm-diameter steel die at 31 MPa, and then pressed hydrostatically at 124 MPa. The resulting compacted powder samples were heated to 1600°C for 8 hours to induce foaming and sintering. The atmosphere was ambient air, except for the pre-sintering environment for the AlN-based foams, where argon was used. The size of the resultant porous pellets depended on the porosity, and the diameter was in the range 16–19 mm, and the samples were 2.5–3.5-mm-thick.

Reference materials were prepared in the form of fully densified ceramics, which were prepared using the same matrix materials. Conventional porous ceramics were also prepared, and the porosity was controlled by adding different quantities of carbon as the pore-forming agent. The heating scheme used for the composite granules was also used for these reference materials.

Five kinds of 8-mol% yttria-stabilized zirconia ceramics were fabricated. The samples were labelled based on the starting matrix powder and foaming agent or pore former. These are summarized in Table 1.

Table 1 Sample labeling based on the matrix, stabilization oxide/superplasticity facilitator, and foaming agent/pore former.

Type	Label	Matrix	Stabilization oxide /plasticity facilitator	Foaming agent /pore former
Conventional	8YSZ-0Y	8YSZ	---	---
	8YSZ-0Y/C	8YSZ	---	/graphite
Superplastically foamed	8YSZ-0Y-silica/SiC	8YSZ	/silica	SiC/
	0YSZ-8Y/SiC	0YSZ	8mol% yttria/	SiC/
	0YSZ-8Y/AlN	0YSZ	8mol% yttria/	AlN/

The apparent density was measured using the Archimedes method with water as the medium. Cross sections were observed using optical and scanning electron microscopy (SEM) (S-4300, Hitachi, Tokyo, Japan). Crystalline phases were identified using X-ray diffraction (XRD) (Multiflex, Rigaku, Tokyo, Japan).

The thermal expansion properties of the materials were characterised using thermomechanical analysis (TMA) (TMA-50, Shimadzu, Kyoto, Japan). The Young's modulus was determined from the flexure–load (δ-P) relation, which was obtained by bending a rectangular sample bar (0.5 mm × 3 mm × 15 mm). Flexure was calibrated using a standard sample that was 4.5 times thicker than the test piece. The load per unit flexure was obtained from the gradient of the δ–P curve. The Young's modulus was calculated as follows:

$$E = 1/4(L^3/t^3 w)(P/\delta), \qquad (1)$$

where L is the span in the bending test, t is the thickness of the specimen, and w is the width of the specimen. The thermal diffusivity was measured using the laser flash method (LFA447 NanoFlash, Netzch, Japan), and was used to calculate the thermal conductivity, specific heat and density.

RESULTS AND DISCUSSION

3.1 SiC-based superplastically foamed ceramics

Post-stabilization was applied to the 8YSZ-based superplastically foamed ceramics, providing closed pores inclusive porous ceramics (0YSZ-8Y/SiC) with a porosity as high as 40%. We have previously reported that superplastically foaming could not be realized using previously stabilized 8YSZ powder without silica.[5]

For comparison, we also fabricated silica-dispersed superplastically foamed ceramics (8YSZ-0Y-silica/SiC) as well as conventionally fabricated porous ceramics using graphite as the pore former (8YSZ-0Y/C). Figure 1 shows a comparison of the elastic moduli with the same porosity (i.e., 40%).

Both superplastically foamed porous ceramics exhibited elastic moduli that were more than twice as large as those of the conventionally fabricated porous ceramics. Compared with the silica-dispersed 8YSZ (i.e., 8YSZ-0Y-silica/SiC), the elastic modulus of the 8YSZ-based superplastically foamed porous ceramic (0YSZ-8Y/SiC) was larger by 20%. This is attributed to the dispersed silica existing as an amorphous phase, with a relatively small elastic modulus.

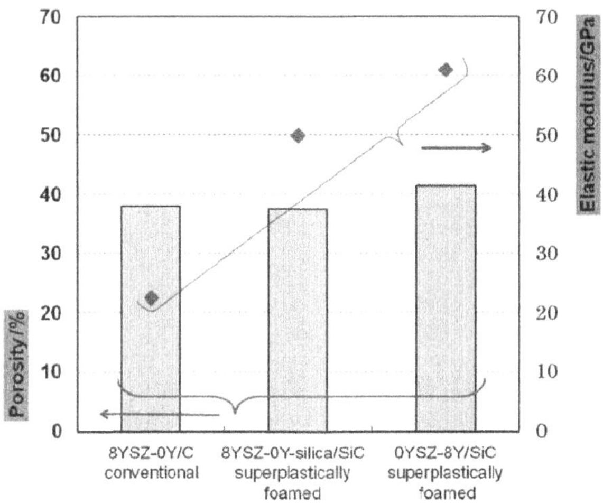

Figure 1. Elastic modulus of conventional and superplastically foamed porous 8YSZ based ceramics. In the superplastically foamed ceramics, silica or 8mol% of yttria was dispersed as superplasticity facilitator.

3-2 Thermal expansion coefficient

A small coefficient of thermal expansion (CTE) is desirable to reduce cyclic fatigue associated with heating and cooling cycles. Because the CTE of 8YSZ is smaller than that of 3YSZ, 8YSZ-based materials are favourable for high-temperature processing and/or high-temperature applications [18].

The linear CTE of the superplastically foamed 8YSZ is shown in Fig. 2. The CTEs for the dense and conventionally fabricated porous 8YSZ ceramics are also shown for comparison. The silica-containing porous material (8YSZ-0Y-silica/SiC) exhibited a significantly larger CTE than the other materials, which is ascribed to the presence of silica.

Comparing the 8YSZ-based ceramics without silica, an anomalous increase in the CTE was observed at around 300°C in the 0YSZ-8Y/SiC. This anomaly was small, but similar to that observed for 8YSZ-0Y-silica/SiC. Since silica was not (intentionally) added to the 0YSZ-8Y/SiC; it appears that the foaming agent SiC formed silica, which led to the CTE anomaly.

To investigate whether the CTE anomaly was related to the SiC foaming agent, we changed the foaming agent to AlN.[13,25] Figure 3 shows a comparison of the CTE of the AlN-incorporated post-stabilized superplastically foamed ceramic (0YSZ-8Y/AlN) with those of 0YSZ-8Y/SiC and the dense 8YSZ ceramics. The anomalous increase in the CTE at around 300°C was not observed for 0YSZ-8Y/AlN, and the CTE profile of 0YSZ-8Y/AlN was similar to that of the dense 8YSZ ceramics.

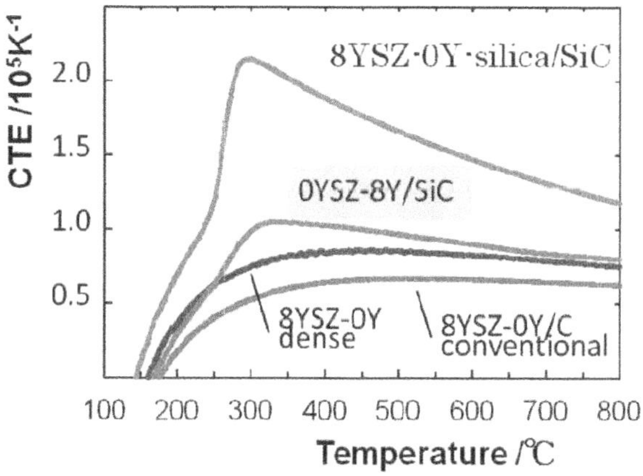

Figure 2. Linear thermal expansion coefficient (CTE) for 8YSZ based dense and porous ceramics

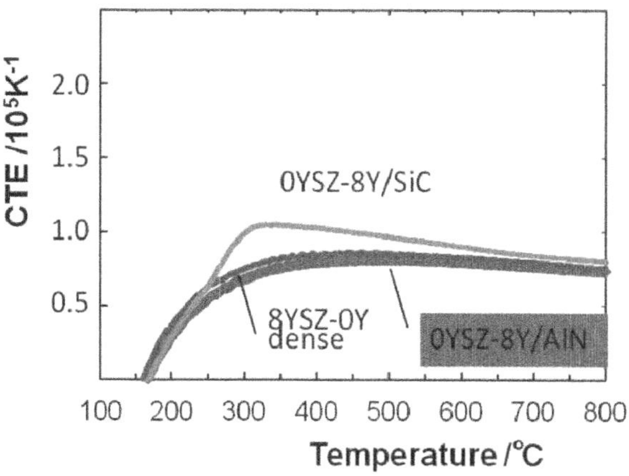

Figure 3. Linear thermal expansion coefficient (CTE) for superplastically foamed porous ceramics using AlN as a foaming agent.

3-3 Thermal conductivity

The thermal conductivities of both the porous ceramics 0YSZ-8Y/SiC and 0YSZ-8Y/AlN were characterised using the laser flash method. With the same porosity (i.e., ~ 40%), we found 1.21 $Wm^{-1}K^{-1}$ for 0YSZ-8Y/SiC and 2.12 $Wm^{-1}K^{-1}$for 0YSZ-8Y/AlN. The latter value coincides with that for the fully dense ceramic (i.e., 2.15 $Wm^{-1}K^{-1}$) despite the significantly differing porosity.

The thermal conductivity of a porous ceramics can be described using the mixture rule,[18] which states that, for a given matrix composition, the porous ceramic will exhibit a lower thermal conductivity than the dense material.

We have already shown that AlN reacts with the ambient oxygen to form alumina prior to gas-evolvable oxidation, leading to pore expansion[13]. In the samples reported here, we used XRD analysis to investigate the formation of alumina in 0YSZ-8Y/AlN. Alumina is expected to contribute to thermal conduction, even in trace amounts, because of its large thermal conductivity (approximately fifteen times that of the 8YSZ ceramics).

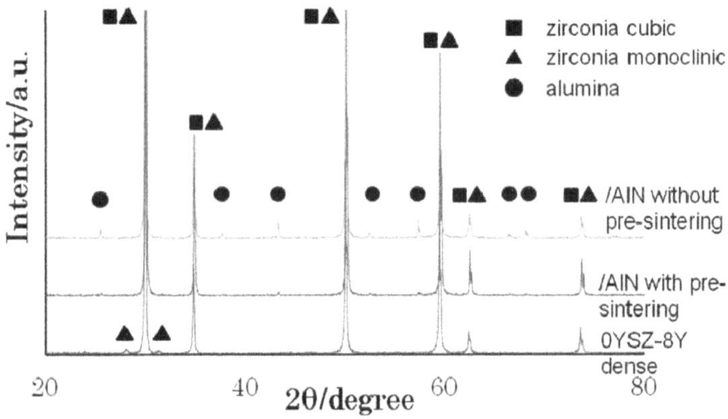

Fig.4 XRD for variously fabricated 0YSZ-8Y ceramics dense and superplastically foamed ceramics with and without pre-sintering

3-4 Modification of the heating program

The thermal conductivity of 0YSZ-8Y/AlN can be reduced to that of 0YSZ-8Y/SiC if the formation of alumina can be suppressed. To prevent the oxidation of AlN prior to sintering/foaming, we modified the heating environment. Prior to the sintering/foaming process, the matrix/foaming agent powder compact was pre-sintered at 1200°C or 1400°C in an argon atmosphere.

Figure 4 shows the XRD profile of 0YSZ-8Y/AlN ceramics with and without pre-sintering. No characteristic monoclinic peaks were observed in the pre-sintered samples, and almost all of the zirconia peaks can be assigned to the cubic phase. Clear alumina peaks were observed in the 0YSZ-8Y/AlN without pre-sintering, which were absent in the sample that had been pre-sintered. This indicates that the formation of an alumina phase prior to the sintering/foaming process was suppressed via the pre-sintering process in an argon environment.

Figure 5 shows the dependence of the porosity of the 0YSZ-8Y/AlN ceramics on the quantity of foaming agent used, with pre-sintering at 1200°C and

1400°C, as well as without pre-sintering. The porosity increased gradually with the quantity of AlN foaming agent, and saturated at around 33% when the AlN content reached the range 15–20 wt%.

When sintering/foaming was carried out following pre-sintering, the dependence of the porosity on the quantity of foaming agent was large. When the quantity of foaming agent was 5 wt%, the porosity was 20% without pre-sintering, and it was 25% with pre-sintering at 1200°C, and 35% with pre-sintering at 1400°C.

The increase in the porosity due to pre-sintering can be attributed to the inhibition of the formation of alumina by eliminating the oxygen supply from the consolidated matrix during the pre-sintering process in the argon atmosphere. This reduction in the formation of alumina was confirmed using XRD measurements.

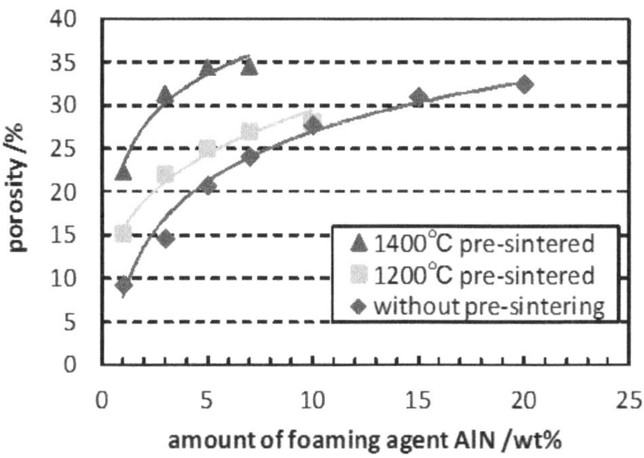

Figure 5. AlN amount dependence of porosity in 0YSZ8Y/AlN with and without pre-sintering

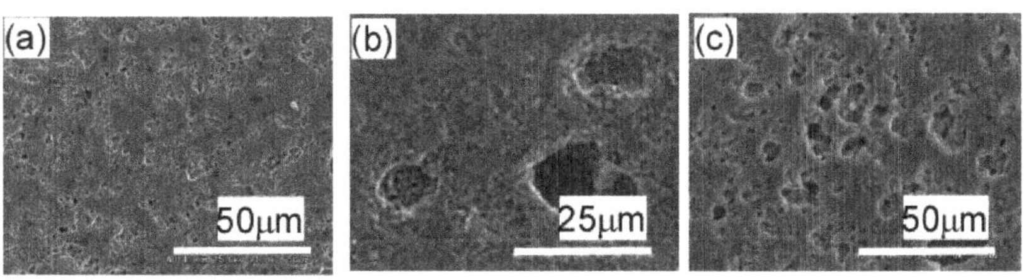

Figure 6. SEM photos of 0YSZ-8Y/AlN with and without pre-sintering.(a)3wt% AlN, without pre-sintering (b)20wt% AlN, without pre-sintering and (c) 3wt% AlN, with pre-sintering

Table 2 Pore sizes and thermal conductivities of 0YSZ-8Y/AlN with and without pre-sintering.

	Low AlN /without pre-sintering	High AlN /withou pre-sintering	Pre-sintered at 1400°C
Quantity of AlN /wt%	3	20	3
Porosity /%	15.1	31.1	33.2
Average pore size / m	4.2	22.4	8.0
Thermal conductivity /$Wm^{-1}K^{-1}$	1.42	2.21	1.19

Table 2 lists the pore sizes and thermal conductivities of the samples with and without pre-sintering, and Figure 6 shows the microstructure. When the quantity of AlN foaming agent AlN was 3 wt%, the porosity of the pre-sintered 0YSZ-8Y/AlN was almost twice that without pre-sintering. Without pre-sintering, more than six times as much AlN (i.e., 20 wt%) was required to obtain the same porosity as with pre-sintering at 1400°C. With 20 wt% of AlN, the pore size was large (i.e., 20 m); however, the pore size remained relatively small with pre-sintering at 1400°C.

With 20 wt% of AlN and without pre-sintering, the pore size was considerably larger than for the sample with 3 wt% of AlN that was pre-sintered at 1400°C, whereas the porosities of the two materials were similar. If the pores expanded with a constant number density, the porosity would increase in proportion to the pore volume; however, as shown in Fig. 6, the number density decreased due to coagulation of the pores.

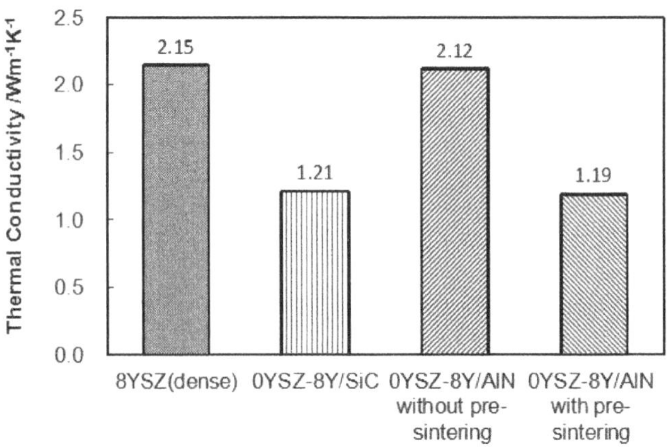

Figure 7. Thermal conductivity of 8YSZ base dense and porous ceramics

Note also that the thermal conductivity of the high-AlN-content 0YSZ-8Y/AlN was larger than that of the low-AlN-content sample, although the porosity of the latter was almost twice that of the former. As discussed previously, this can be attributed to the formation of alumina, which has a large thermal conductivity.

Figure 7 shows the thermal conductivities of the superplastically foamed 8YSZ-based ceramics. The thermal conductivity of 0YSZ-8Y/SiC was approximately half of the dense 8YSZ ceramics; however, it exhibited anomalous thermal expansion due to the silica (which was derived from the SiC foaming agent). This anomalous thermal expansion was eliminated by changing the foaming agent from SiC to AlN. The thermal conductivity of the 0YSZ-8Y/AlN foamed ceramic, however, was as large as that of the dense material, irrespective of the porous nature.

It appears that an alumina phase, formed prior to the sintering/foaming process, leads to an increase in the thermal conductivity. To inhibit the formation of alumina, a pre-sintering process was carried out in an argon atmosphere prior to the sintering/foaming process. This allowed us to obtain foamed ceramics with a comparable porosity using only a small quantity of AlN.

By combining the use of the AlN foaming agent and the pre-sintering process in an argon atmosphere, we obtained ceramics with a thermal conductivity as small as that of 0YSZ-8Y/SiC; however, the 0YSZ-8Y/AlN foamed ceramic is favorable for high-temperature applications because it does not exhibit the anomalous thermal expansion.

CONCLUSIONS

Our innovated superplastically foaming method was adopted to make closed-pore inclusive zirconia-based ceramics. Yttrium oxide was added to monoclinic yttria-stabilized zirconia to make full stabilization of the matrix. Sintering and superplastic deformation led to a solid solution and transformation to the cubic phase. Such porous ceramics through post-stabilization method showed anomalous thermal expansion due to the silica derived from the foaming agent SiC. Such anomalous thermal expansion disappeared by changing the foaming agent from SiC to AlN. The resulting superplastically foamed porous ceramics showed relatively large thermal conductivity comparable to that of the fully densified ceramics while the porosity was as large as 40%. This high thermal conductivity was due to the alumina derived from the foaming agent AlN. To suppress the formation of alumina, the powder compacts was pre-sintered in argon atmosphere. The modified superplastically foamed ceramics had smaller pores with porosity of 33%, which shows favorable thermal insulation almost half thermal conductivity as that of the full dense one.

Acknowledgements
A part of this work has been supported by JSPS Grant-in-Aid for Scientific Research (B) 25289229.

REFERENCES

[1]Kishimoto, A., Higashiwada, T., Asaoka, H. and Hayashi, H. "The exploitation of superplasticity in the successful foaming of ceramics following sintering",Adv. Eng. Mater. (8) pp.708-11,2006.
[2]Kishimoto, A., Higashiwada, T., Takahara, M. and Hayashi, H. "Solid state foaming and free-forming of closed pore utilizing the superplasticity of zirconia ceramics",Mater. Sci. Forum (544-54 544-545) pp.641-4,2007.

[3]Higashiwada, T., Asaoka, H., Hayashi, H. and Kishimoto, A. "Effect of additives on the pore evolution of zirconia based ceramic foams after sintering",J. Eur. Ceram. Soc. (27) pp.2217-22,2007.
[4]Kishimoto, A., Obata, M. and Hayashi, H. "Fabrication of alumina-based ceramic foams utilizing superplasticity",J. Eur. Ceram. Soc. (27) pp.41-5,2007.
[5]Hashida, Y., Hayashi, H. and Kishimoto, A. "Fabrication of Solid State Foams Based on Full Stabilized Zirconia Ceramics Facilitating the Superplasticity with Dispersoids(in Japanese)",J. Jpn. Soc. Powder Powder Metallurgy (54) pp.740-3,2007.
[6]Hanao, M., Hayashi, H. and Kishimoto, A. "The mechanical and thermal properties of porous zirconia ceramics fabricated through a solid-state foaming method",J. Jpn. Soc. Powder Powder Metallurgy (54) pp.732-7,2008.
[7]Kishimoto, A., Obata, M., Waku, K. and Hayashi, H. "Mechanical and electrical properties of superplastically foamed titania based ceramics",Ceram. Int. (35) pp.1441-5,2009.
[8]Obata, M., Kishimoto, A. and Hayashi, H. "Comparison of magnesia and magunesium alumina spinel on the alumina based superprastisity foam",J. Alloys. Compd. (471) pp.32-5,2009.
[9]Kishimoto, A., Hanao, M. and Hayashi, H. "Improvement in the specific strength by arranging closed pores in fully densified zirconia ceramics",Adv. Eng. Mater. (11) pp.96-100,2009.
[10]Nishino, Y., Hayashi, H. and Kishimoto, A. "Influence of processing parameters on the closed pore patterns utilizing the superplastisically foaming method (in Japanese)",J. Jpn. Soc. Powder Powder Metallurgy (56) pp.389-94,2009.
[11]Wakiyama, M., Waku, K., Hayashi, H. and Kishimoto, A. "Fabrication of closed-pore inclusive low-permittivity substrates insensitive to ambient humidity",J. Ceram. Soc. Jpn. (117) pp.1013-6,2009.
[12]Kishimoto, A. "Superplasticity in ceramics - A new application",J. Tech. Assoc. Refract. Jpn. (29) pp.3-7, 2009.
[13]Yamaoka, H., Hayashi, H. and Kishimoto, A. "The applicability of nitride powders as foaming agents in superplastically foamed ceramics",J. Ceram. Soc. Jpn. (117) pp.1233-5, 2009.
[14]Kishimoto, A., Nishino, Y. and Hayashi, H. "Patterning of closed pores utilizing the superplastically foaming method",Ceram. Trans. (212) pp.143-51, 2010.
[15]Wakiyama, M., Hayashi, H. and Kishimoto, A. "Alumina based low permitivity substrate utilizing superplastically foaming method",J. Ceram. Soc. Jpn. (118) pp.1090-3, 2010.
[16]Kishimoto, A., Wakiayma, M. and Hayashi, H. "Superplastically foaming method to make closed pores inclusive alumina based ceramics",World J. Eng. (7) pp.P536. , 2010.
[17]Bhuiyan, T., Hayashi, H. and Kishimoto, A. "Formation of 3D tubular canal in fully densified zirconia ceramics by free foaming method using superplasticity",Ceram. Int. (36) pp.399-402, 2011.
[18]Kishimoto, A., Nakagawa, T., Teranishi, T. and Hayashi, H. "Superplastically foamming method for reliable porous ceramics",Mater. Sci. Forum (735) pp.109-12, 2013.
[19]Kishimoto, A., Okada, M., Teranishi, T. and Hayashi, H. "Maintaining the mechanical strength of La-, Y- co-substituted zirconia porous ceramics through the superplastically foaming method",Mater. Sci. & Eng. A (581) pp.98-103, 2013.

[20]Marushima, T., Goto, T., Hirai, T. and Iguchi, Y. "High-Temperature Oxidation of Silicon Carbide and Silicon Nitride",Mater. Trans. JIM (38) pp.821-35, 1997.
[21]Klemm, H. and Hermann, H. "High Temperature Properties of Si3N4 Materials",J. Eur. Ceram. Soc. (7) pp.315-8, 1991.
[22]Hinze, J.W. and Graham, H.C. "The Active Oxidaiotn of Si and SiC in the Viscous Gas-Flow Regime",Journal of the Electrochemical Society (123) pp.1066-73,1976.
[23]Antill, J.E. and Warburton, J.B. "Active to passive transition in the oxidation of SiC",Corros. Sci. (11) pp.337-42, 1971.
[24]Narushima, T., Goto, T., Iguchi, Y. and Hirai, T. "High-Temperature Active Oxidaiton of Chemically Vapor-Deposited Silicon Carbide in an Ar-O2 Atmosphere",J. Am. Ceram. Soc. (74) pp.2583-6, 1991.
[25]Fan, Z.Y. and Newman, N. "Experimental determination of the rates of decomposition and cation desorption from AlN surfaces",Mater. Sci. & Eng. B (87) pp.224-48, 2001.

The English in this document has been checked by at least two professional editors, both native speakers of English. For a certificate, please see: http://www.textcheck.com/certificate/tYuBpP

CYANOSILYLATION OF BENZALDEHYDE WITH TRIMETHYLSILYL CYANIDE OVER A-SITE METAL SUBSTITUTED PEROVSKITE-TYPE OXIDE CATALYST PREPARED BY THERMAL DECOMPOSITION OF HETERONUCLEAR CYANO COMPLEX PRECURSORS

Syuhei Yamaguchi, Hiroki Wada, Takahisa Okuwa, Hidenori Yahiro*
Department of Materials Science and Biotechnology, Graduate School of Science and Engineering, Ehime University
3 Bunkyo-cho, Matsuyama, Japan

ABSTRACT

The perovskite-type oxide catalyst, $LnFeO_3$ (Ln = La, Pr, Sm, Gd, and Dy), prepared by decomposition method of heteronuclear cyano complex precursor (CN method) was applied to liquid-phase organic reaction, cyanosilylation of benzaldehyde with TMSCN. The catalytic activity per specific surface area over $LaFeO_3$ was dependent on calcination temperature in CN method and it was found that the catalytic activity was strongly related to crystallinity of catalyst. The catalytic activity of $LnFeO_3$ catalyst increased with decreasing ionic radius of A-site lanthanoid ion. The catalytic activity of $LnFeO_3$ catalyst was found to increase with increasing surface oxygen concentration on $LnFeO_3$, which was estimated by XPS results.

INTRODUCTION

Perovskite-type oxide, ABO_3, exhibited high catalytic activity in oxidation process and high thermal stability. Up to date, numerous efforts of preparing perovskite-type oxides with high surface area have been devoted. Recently, low temperature wet-chemical methods such as sol-gel[1-4], co-precipitation[4,5], reverse homogeneous precipitation (RHP)[6], and polymeric precursor (PP) methods[7], have been reported to prepare nano-sized perovskite-type oxides. A preparation route of perovskite-type oxide has been also reported via the thermal decomposition of heteronuclear cyano complex, $M[M'(CN)_6]\cdot nH_2O$ (abbreviated as CN method), where M and M' stand for lanthanide and transition metal ions, respectively[8-15]. Several perovskite-type oxides prepared by low temperature wet-chemical methods have been reported to show the high catalytic activity for catalytic reaction in gas phase, such as the complete oxidation of hydrocarbons[16] and CO[17] and the decomposition of NO[18,19]. However, little is known about the catalytic activity of perovskite-type oxide for the organic reaction in liquid phase. Recently, we reported that the perovskite-type oxide catalyst, $SmFeO_3$, prepared by CN method was active for liquid-phase organic reaction, cyanosilylation of benzaldehyde with trimethylsilyl cyanide (TMSCN)[20]. It was clear that the catalytic activity per specific surface area over $SmFeO_3$ catalyst prepared by CN method was higher than those by PP and RHP methods. In the present paper, we wish to report the catalytic activity for cyanosilylation of carbonyl compounds with TMSCN over A-site metal substituted perovskite-type oxide,

$LnFeO_3$ (Ln = La, Pr, Sm, Gd, and Dy), prepared by CN method. The cyanosilylation of carbonyl compounds with trimethylsilyl cyanide (TMSCN) is one of very important transformations to produce cyanohydrin trimethylsilyl ethers which can be further transformed into important synthetic intermediates such as α-hydroxy acids, α-amino acids, and β-amino alcohols[21,22]. Many researchers have already reported this transformation using several catalysts with Brønsted[23] acid, Lewis acid[24,25], and Lewis bases[26].

EXPERIMENTAL

The heteronuclear cyano complex, $La[Fe(CN)_6]{\bullet}5H_2O$, as a precursor of perovskite-type oxide was synthesized at room temperature as reported elsewhere. $La(NO_3)_3{\cdot}6H_2O$ (Wako, 99.5%) and $K_3Fe(CN)_6$ (Hayashi, 99.0%) were used as starting materials. The $La[Fe(CN)_6]{\bullet}5H_2O$ precursor was calcined at 500–1000 °C for 1 h in air. The other heteronuclear cyano complex precursors, $Ln[Fe(CN)_6]{\bullet}nH_2O$ (Ln = Pr, Sm, Gd, and Dy), were prepared by the same method. The obtained precursors were calcined at 1000 °C for 1 h in air.

The crystalline structure of samples was measured by the powder XRD (Rint 2000, Rigaku) with Cu-K radiation. Ln and Fe contents in samples were measured by X-ray fluorescence spectrometer (RIX2100 (Rigaku), Rh K radiation source). The specific surface area was determined with the BET analysis (Belsorp-mini, BEL Japan) for the adsorption–desorption measurements using N_2 adsorbent at 77 K. The surface chemical states and compositions of perovskite-type oxides were evaluated by X-ray photoelectron spectroscopy (XPS) analysis (XPS 1600E, PerkinElmer) using an MgK excitation source. All the binding energy (BE) values in XPS spectra were referenced to the C 1s line at 284.5 eV. In addition, relative atomic sensitivity factors (ASF) were used to determine practically more accurate

Table 1 Characteristic data of catalysts used in the present study.

Sample	Calcination condition	Surface area/ $m^2\ g^{-1}$	Crystallite size/ nm [a]	Atomic composition [b]
	Temp./ °C			Sm/Fe
$La[Fe(CN)_6]{\bullet}5H_2O$	—	—	—	0.90
$LaFeO_3$	500	17	—	0.95
$LaFeO_3$	550	28	21	0.97
$LaFeO_3$	600	18	25	0.95
$LaFeO_3$	700	9.8	29	0.95
$LaFeO_3$	800	4.8	35	0.95
$LaFeO_3$	900	4.7	38	0.97
$LaFeO_3$	1000	3.0	40	0.95
$PrFeO_3$	1000	1.8	44	0.96
$SmFeO_3$	1000	1.5	40	0.97
$GdFeO_3$	1000	1.6	37	0.93
$DyFeO_3$	1000	1.1	43	0.97

a) The average crystallite size in the (110) direction was evaluated from XRD line broading using Scherrer's equation. b) XRF analysis.

chemical composition on the surface. The samples used in the present study are listed in Table 1.

Benzaldehyde (Wako, 98.0%), trimethylsilyl cyanide (TMSCN) (TCI, 98.0%), and -(trimethylsilyl)phenylacetonitrile (Sigma Aldrich, 98%) were used without any purification. A typical procedure for cyanosilylation of benzaldehyde is as follows; catalyst (50 mg), benzaldehyde (0.5 mmol), and dichloromethane (12 mL) were successively placed into a Pyrex-glass test tube. The reaction was initiated by the addition of TMSCN (1.0 mmol) and the reaction mixture was vigorously stirred (800 rpm) at 20 °C in air (1 atm). The conversion of benzaldehyde and the yield of product, -(trimethylsilyl)phenylacetonitrile, were periodically determined by GC analysis.

RESULTS AND DISCUSSION

XRD profiles of the $La[Fe(CN)_6]•5H_2O$ precursors calcined at 500-1000 °C for 1 h were depicted in Figs. 1(b)-1(h). The XRD profile of as-prepared $La[Fe(CN)_6]•5H_2O$ precursor (JCPDS# = 25-1198) is also shown in Fig. 1(a). The $La[Fe(CN)_6]•5H_2O$ precursor calcined at 500 °C showed no intense XRD peak, suggesting that an amorphous phase was formed (Fig. 1(b)). The samples calcined at 550-1000 °C provided the XRD pattern indicating the single phase of the perovskite-type structure of $LaFeO_3$ (JCPDS# = 37-1493) (Fig. 1(c-h)). The XRD peak intensities of $LaFeO_3$ increased with increasing calcination temperature; the crystallite diameter of $LaFeO_3$, which was determined by Scherrer's equation, decreased as follows: 1000 (40 nm) > 900 (38) > 800 (35) > 700 (29) > 600 (25) > 550 °C (21). A similar result was reported for $SmFeO_3$ prepared by CN method[26]. However, the lowest calcination temperature (550 °C) at which the single phase of $LaFeO_3$ was observed was lower than that of $SmFeO_3$ (700 °C)[20], indicating that the transformation of $La[Fe(CN)_6]•5H_2O$ to $LaFeO_3$ easier proceeds

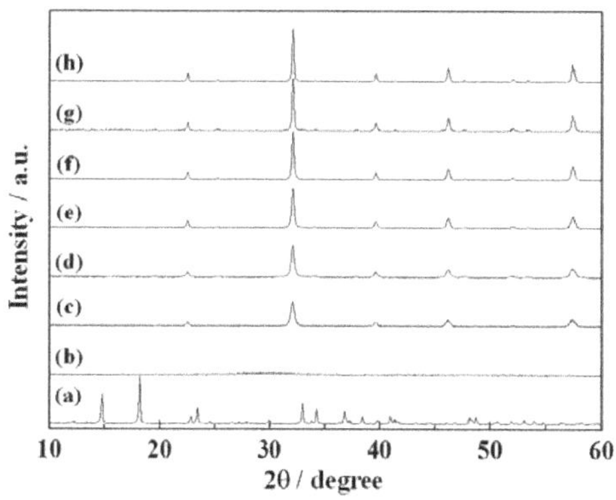

Figure 1. XRD patterns of LaFe-containing samples. A heteronuclear metal cyano complex precursor, $La[Fe(CN)_6]•5H_2O$ (a), was calcined at 500 (b), 550 (c), 600 (d), 700 (e), 800 (f), 900 (g), and 1000 °C (h) for 1 h in air.

than that of $La[Fe(CN)_6]•5H_2O$ to $SmFeO_3$. This result may be related to the geometrical factor that $LaFeO_3$ is less distorted in crystal structure than $SmFeO_3$, because Goldschmidt tolerance factor (t) of $LaFeO_3$ (t = 0.926) was higher than that of $SmFeO_3$ (t = 0.897)[27].

Figures 2(A) and 2(B) show catalytic activities per catalyst weight and per specific surface area, respectively, for cyanosilylation of benzaldehyde with TMSCN over $LaFeO_3$ catalysts calcined at several temperatures. As shown in Fig. 2(A), the catalytic activity per

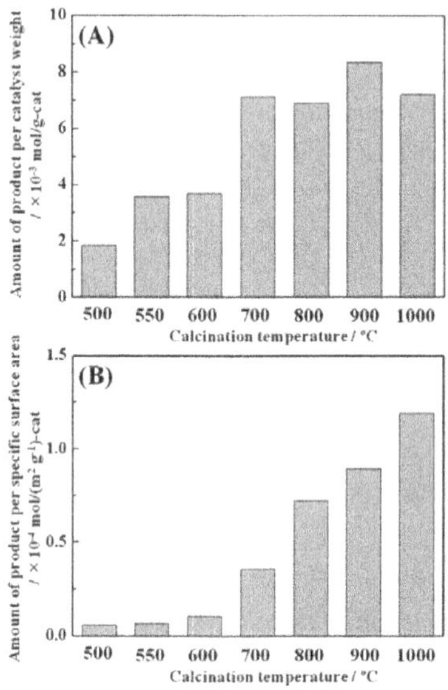

Figure 2. Catalytic activities (A) per catalyst weight and (B) per specific surface area over $LaFeO_3$ catalysts calcined at several temperatures. Reaction conditions: benzaldehyde (0.5 mmol), TMSCN (1 mmol), catalyst (50 mg), CH_2Cl_2 (12 ml), 20 °C, and 1 h.

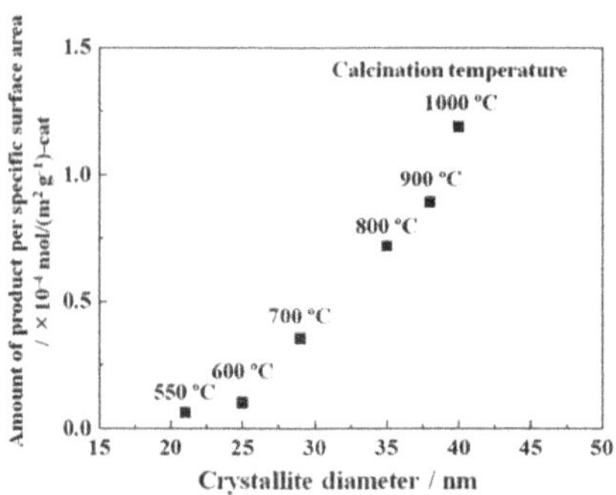

Figure 3. Catalytic activities for cyanosilylation as a function of crystallite diameter of $LaFeO_3$ catalysts. Reaction conditions: benzaldehyde (0.5 mmol), TMSCN (1 mmol), catalyst (50 mg), CH_2Cl_2 (12 ml), 20 °C, and 1 h.

catalyst weight increased with increasing calcination temperature, reached the maximum value at 900 °C, then slightly decreased with further increasing calcination temperature. On the other hand, Figure 2(B) shows that the catalytic activity per specific surface area monotonously increases with increasing calcination temperature. We have reported that the catalytic activity per specific surface area for cyanosilylation increased with increasing calcination temperature for $SmFeO_3$ catalyst prepared by CN method and that the catalytic activity per specific surface area was correlated with the crystallite diameter of $SmFeO_3$ catalysts[20]. Figure 3 shows the catalytic activities per specific surface area as a function of the crystallite diameter of $LaFeO_3$ catalysts, estimated from XRD (Table 1). It is clear that the catalytic activity increased with increasing crystallite diameter, suggesting that the crystal growth with the long-term ordering of perovskite-type structure, plays an important role in increasing the catalytic activity for cyanosilylation of benzaldehyde with TMSCN, similar to the case of $SmFeO_3$ system[20].

Next, we investigated the effect of A-site cation on the catalytic property for cyanosilylation. Figure 4 shows XRD profiles of $Ln[Fe(CN)_6]\cdot nH_2O$ calcined at 1000 ˚C for 1 h to yield $LnFeO_3$. A single phase of perovskite-type oxide was obtained at 1000 ˚C for Ln = La, Pr, Sm, Gd, and Dy. The crystal structure of perovskite-type oxide was in rhombohedral for Ln = La, in cubic for Pr, and in orthorhombic for Sm–Dy. A similar result has been reported in $LnCoO_3$ system[28]. Figure 5 shows catalytic activities per catalyst weight for cyanosilylation of benzaldehyde with TMSCN over $LnFeO_3$ calcined at 1000 ˚C for 1 h. The catalytic activity of $LnFeO_3$ catalyst increased in the following order: Ln = La $\approx$ Pr $<$ Sm $\approx$ Gd $<$ Dy. The catalytic activity per specific surface area tended to increase with decreasing ionic radius of A-site lanthanoid ion Ln^{3+}. Therefore, the catalytic activity is probably related to the distorting of the crystal structure of perovskite-type oxide in $LnFeO_3$ because of the decrease in Goldschmidt tolerance factor of $LnFeO_3$ with increasing ion radius.

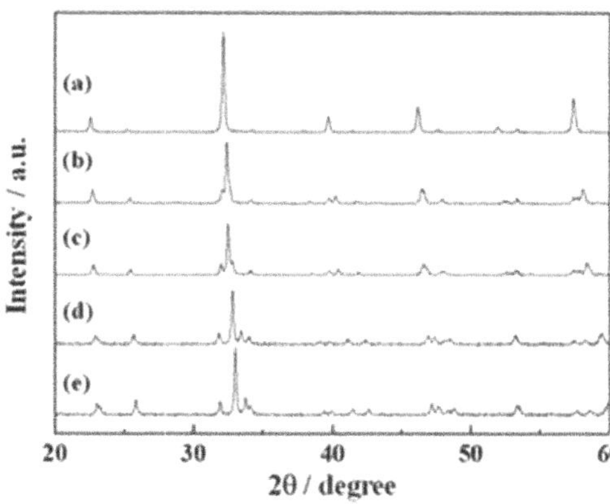

Figure 4. XRD patterns of LnFe-containing samples calcined at 1000 ˚C for 1 h in air. (a) $LaFeO_3$, (b) $PrFeO_3$, (c) $SmFeO_3$, (d) $GdFeO_3$, and (e) $DyFeO_3$.

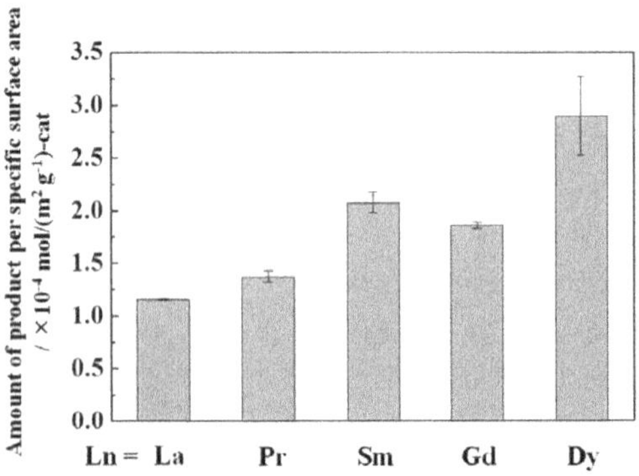

Figure 5. Catalytic activities per specific surface area over $LnFeO_3$ catalysts (Ln = La, Pr, Sm, Gd, and Dy) calcined at 1000 °C for 1 h in air. Reaction conditions: benzaldehyde (0.5 mmol), TMSCN (1 mmol), catalyst (50 mg), CH_2Cl_2 (12 ml), 20 °C, and 1 h.

XPS analyses were carried out to investigate the surface state of $LnFeO_3$ prepared by CN method. XPS results are summarized in Table 2. We focused here on Fe 2p3/2 and O1s XPS peaks observed at 710.0 ± 0.5 eV and 531.0 ± 0.2 eV, which can be assigned to Fe^{3+} and surface oxygen, respectively[11]. The binding energies of Fe^{3+} and surface oxygen in the samples calcined at 1000 °C for 1 h were less affected by the kind of lanthanoid ion in $LnFeO_3$, while the intensities of these peaks were affected. The ratio of XPS peak area of surface oxygen to that of whole cations in $LnFeO_3$, Rosurface, was defined by

$$R_{O_{surface}} = \frac{A_{O_{surface}}/F_{O_{surface}}}{(A_{Ln}/F_{Ln}) + (A_{Fe}/F_{Fe})}$$

where A_{Ln}, A_{Fe}, and Aosurface are XPS peak areas of Ln^{3+}, Fe^{3+}, and surface oxygen, respectively, and F_{Ln}, F_{Fe}, and Fosurface are ASF values of Ln, Fe, and O elements, respectively. The Rosurface values for each samples are listed in Table 2. Rosurface increased in the following order: La < Pr < Sm < Gd < Dy. Figure 6 shows the catalytic activity per specific

Table 2 XPS binding energy and $R_{O_{surface}}$ of $LnFeO_3$ catalysts

Ln	Binding energy/ eV				$R_{O_{surface}}$
	Fe 2p3/2	Fe 2p1/2	$O1s_{lattice}$	$O1s_{surface}$	
La	709.6	723.0	529.0	531.2	0.15
Pr	710.2	723.6	528.9	531.1	0.49
Sm	709.9	723.4	528.6	531.0	1.06
Gd	710.1	723.6	528.8	530.9	1.34
Dy	709.9	723.7	528.8	530.9	1.73

surface area as a function of Rosurface value in $LnFeO_3$ catalyst. Catalytic activity increased with increasing Rosurface value of $LnFeO_3$. Karpova et al. reported that the amount of Brønsted acid sites increased with increasing surface oxygen concentration[29]. In our paper reported previously[20], it was shown that the catalytically active sites are the Brønsted acid sites formed on perovskite-type oxide catalyst. Considering above reports, the increase in the ionic radius of A-site cation may cause the increase in the amount of Brønsted acid sites, resulting in the increase in the catalytic activity for cyanosilylation of benzaldehyde with TMSCN.

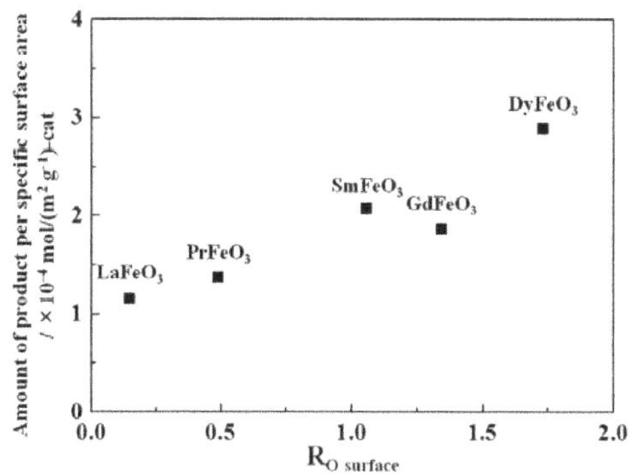

Figure 6. Catalytic activities per specific surface area as a function of surface oxygen concentration on $LnFeO_3$ catalysts (Ln = La, Pr, Sm, Gd, and Dy) calcined at 1000 °C for 1 h in air (CN method). Reaction conditions: benzaldehyde (0.5 mmol), TMSCN (1 mmol), catalyst (50 mg), CH_2Cl_2 (12 ml), 20 °C, and

CONCLUSIONS

As a model organic reaction in liquid phase, the cyanosilylation of benzaldehyde with TMSCN was performed over perovskite-type oxide, $LnFeO_3$. For $LaFeO_3$ catalyst prepared by CN method, the catalytic activity per catalyst weight increased with increasing calcination temperature of $La[Fe(CN)_6]$ complex precursor, reached maximum value at 900 °C, and slightly decreased with further increasing calcination temperature. Taking XRD results into consideration, it was found that the catalytic activity was strongly related to crystallinity of $LaFeO_3$ catalyst. For $LnFeO_3$ prepared by CN method, the catalytic activity per specific surface area catalyst increased in the following order: Ln = La < Pr < Sm ≈ Gd < Dy; the catalytic activity is probably related to the distorting of the crystal structure of perovskite-type oxide in $LnFeO_3$. The surface concentration of oxygen, Rosurface, estimated by XPS results, increased with increasing ionic radius of A-site cation: La < Pr < Sm < Gd < Dy. It was found that the catalytic activity of $LnFeO_3$ catalyst tended to increase with increasing surface oxygen concentration on $LnFeO_3$, suggesting that the increase in Brønsted acid sites on the surface of perovskite-type oxide catalyst may enhance the catalytic activity for cyanosilylation of benzaldehyde with TMSCN.

ACKNOWLEDGEMENTS

This work was supported by JSPS KAKENHI Grant Numbers 24560948 and 25820394.

REFERENCES

[1] Zhang, H. M., Teraoka, Y., & Yamazoe, N. (1987). Preparation of perovskite-type oxides with large surface area by citrate process. *Chem. Lett.*, 665-668.

[2] Bell, R. J., Millar, G. J., & Drennan, J. (2000). Influence of synthesis route on catalytic properties of $La_{1-x}Sr_xMnO_3$. *Solid State Ionics 131*, 211-220.

[3] Rida, K., Benabbas, A., Bouremmad, F., Pena, M. A., Sastre, E., & Martine-Arias, A. (2007). Effect of calcination temperature on the structural characteristics and catalytic activity for propene combustion of sol-gel derived lanthanum chromite perovskite. *Appl. Catal. A 327*, 173-179.

[4] Gosavi, P. V., & Biniwale, R. B. (2010). Pure phase $LaFeO_3$ perovskite with improved surface area synthesized using different routes and its characterization. *Mater. Chem. Phys. 119*, 324-329.

[5] Li, W., Wang, N., Wu, W. J., Wang, Y. L., & Zhao, M. Y. (2004). Synthesis of $Ba_2Ti_9O_{20}$ powder via preparation-mixing route. *Mater. Lett. 59*, 162-165.

[6] Teraoka, Y., Nanri, S., Moriguchi, I., Kagawa, S., Shimanoe, K., & Yamazoe, N. (2000). Synthesis of manganite perovskites by reverse homogeneous precipitation method in the presence of alkylammonium cations. *Chem. Lett.*, 1202-1203.

[7] Tsuchida, K., Takase, S., & Shimizu, Y. (2004). Sol-gel synthesis of perovskite-type oxide thin-film susing metal organic acid and its application to amperometric hydrogen-phosphate ion sensor. *Sens. Mater. 16*, 171-180.

[8] Matuura, Y., Matsushima, S., Sakamoto, M., & Sadaoka, Y. (1993). Nitrogen dioxide-sensitive lanthanium iron oxide ($LaFeO_3$) film prepared by thermal decomposition of the heteronuclear complex, lanthanum hexacyanoferrate pentahydrate $(La[Fe(CN)_6]\cdot 5H_2O)_x$. *J. Mater. Chem. 3*, 1329.

[9] Sadaoka, Y., Traversa, E., & Sakamoto, M. (1996). Preparation and characterization of heteronuclear complexes, $Ln_xSm_{1-x}[Co(CN)_6]\cdot nH_2O$ (Ln = La, Er, and Yb). *Chem. Lett.*, 177-178.

[10] Sadaoka, Y., Traversa, E., & Sakamoto, M. (1996). Preparation and structural characterization of perovskite-type $La_xLn''_{1-x}CoO_3$ by the thermal decomposition of heteronuclear complexes, $La_xLn''_{1-x}[Co(CN)_6]\cdot nH_2O$ (Ln" = Sm and Ho). *J. Alloys Compd. 240*, 51-59.

[11] Asamoto, M., Harada, N., Iwamoto, Y., Yamaura, H., Sadaoka, Y., & Yahiro, H. (2009). Catalytic activity of multi-metallic perovskite-type oxide prepared by the thermal decomposition of heteronuclear cyano complex, $Sm[Fe_xCo_{1-x}(CN)_6]\cdot nH_2O$. *Top. Catal. 52*, 823-827.

[12] Asamoto, M., & Yahiro, H. (2009). Catalytic property of perovskite-type oxide prepared by thermal decomposition of heteronuclear complex. *Catal. Surv. Asia 13*, 221-228.

[13] Mori, M., Iwamoto, Y., Asamoto, M., Itagaki, Y., Yahiro, H., Sasaoka, Y., Takase, S., Shimizu, Y., Yuasa, M., Shimanoe, K., Kusaka, H., & Teraoka, Y. (2008). Effect of preparation routes on the catalytic activity over $SmFeO_3$ oxide. *Catal. Today 139*, 125-129.
[14] Asamoto, M., Hino, M., Yamaguchi, S., & Yahiro, H. (2011). Transformation of crystalline heteronuclear cyano complex to crystalline perovskite-type oxide by thermal decomposition. *Catal. Today 175*, 534-540.
[15] Asamoto, M., Iwasaki, Y., Yamaguchi, S., & Yahiro, H. (2012). Synthesis of perovskite-type oxide catalysts, Ln(Fe, Co)O_3 (Ln = La, Pr, Sm, Gd, Dy, Ho, Er, and Yb), from the thermal decomposition of the corresponding cyano complexes. *Catal. Today 185*, 230-235.
[16] Nitadori, T., Ichiki, T., & Misono, M. (1988). Catalytic properties of perovskite-type oxides (ABO_3) consisting of rare earth and 3d transition metals. The roles of the A- and B-site ions. *Bull. Chem. Soc. Jpn. 61*, 621-626.
[17] Mizuno, N., Fujisawa, Y., & Misono, M. (1989). Pronounced synergetic effect in the catalytic properties of $LaMn_{1-x}Cu_xO_3$. *J. Chem. Soc., Chem. Commun.*, 316-318.
[18] Wang, J., Yasuda, H., Inumaru, K., & Misono, M. (1995). Catalytic decomposition of dinitrogen oxide over perovskite-related mixed oxides. *Bull. Chem. Soc. Jpn. 68*, 1226-1231.
[19] Teraoka, Y., Harada, T., & Kagawa, S. (1998). Reaction mechanism of direct decomposition of nitric oxide over Co- and Mn-based perovskite-type oxides. *J. Chem. Soc., Faraday Trans. 94*, 1887-1891.
[20] Yamaguchi, S., Okuwa, T., Wada, H., Yamaura, H., & Yahiro, H. (2015). Cyanosilylation of benzaldehyde with TMSCN over perovskite-type oxide catalyst prepared by thermal decomposition of heteronuclear cyano complex precursors. *Res. Chem. Intermed.* in press. DOI: 10.1007/s11164-015-1980-y.
[21] Gregory, R. J. H. (1999). Cyanohydrins in Nature and the Laboratory: Biology, Preparations, and Synthetic Applications. *Chem. Rev. 99*, 3649-3682.
[22] Ryu, D. H., & Corey, E. J. (2004). Highly Enantioselective Cyanosilylation of Aldehydes Catalyzed by a Chiral Oxazaborolidinium Ion. *J. Am. Chem. Soc. 126*, 8106-8107.
[23] Ogasawara, Y., Uchida, S., Yamaguchi, K., & Mizuno, N. (2009). A Tin-Tungsten Mixed Oxides as an Efficient Heterogeneous Catalyst for C-C Bond-Forming Reactions. *Chem. Eur. J. 15*, 4343-4349.
[24] Kurono, N., Yamaguchi, M., Suzuki, K., & Ohkuma, T. (2005). Lithium chloride: An active and simple catalyst for cyanosilylation of aldehydes and ketones. *J. Org. Chem. 70*, 6530-6532.
[25] Karimi, B., & Ma'Mani, L. (2004). A High Efficient and Recyclable Silica-Based Scandium(III) Interphase Catalyst for Cyanosilylation of Carbonyl Compounds. *Org. Lett. 6*, 4813-4815.
[26] Fetterly, B. M., & Verkade, J. G. (2005). $P(RNCH_2CH_2)_3N$: efficient catalysts for the cyanosilylation of aldehydes and ketones. *Tetrahedron Lett. 46*, 8061-8066.
[27] Liferovich, R. P., & Mitchell, R. H. (2004). A structural study of ternary lanthanide

orthoscandate perovskites. *J. Solid State Chem. 177*, 2188-2197.

28 Parkash, O., Ganguly, P., Rao, G. R., Rao, C. N., Rajoria, D. S., & Bhide, V. G. (1974). Rare earth cobaltite catalysts. Relation of activity to spin and valence states of cobalt. *Mater. Res. Bull. 9*, 1173-1176.

29 Karpova, S. S., Moshnikov, V. A., Maksimov, A. I., Mjakin, S. V., & Kazantseva, N. E. (2013). Study of the effect of the acid-base surface properties of ZnO, Fe_2O_3 and $ZnFe_2O_4$ oxides on their gas sensitivity to ethanol vapor. *Semiconductors 47*, 1026-1030.

Geopolymers, Inorganic Polymer Ceramics, and Sustainable Composites

NANOPARTICLES SEEDED GEOPOLYMERS

Pernechele Matteo [1], Troczynski Tom [1], Pawlik Marek [2]
[1] Material engineering, University of British Columbia, Vancouver, BC, Canada.
[2] Mining Engineering, University of British Columbia, Vancouver, BC, Canada.

ABSTRACT

The use of different nanoparticles as seeding agents for geopolymeric binders was studied to determine the relations between the type of seeds and the geopolymers setting time, structural reorganization, compressive strength and durability. The hypothesis that seeds can act as templates for the reorganization of metakaoline based geopolymers was tested using zeolites with different crystal structures and chemistry, silica and alumina. The strain induced by high energy ball milling on the nanoparticles was determined by X-ray diffraction and by spectroscopic techniques. The effects of microstrains in the seeds on the geopolymeric reaction kinetics and products were investigated. FTIR and X-ray diffraction provided the short order and long order information needed to characterize the structural reorganization of the geopolymer. SEM imaging was adopted to study the effect of the seeds on the microstructure, in particular the nature of the geopolymer product (dense particulates or homogeneous gel). The fresh and hardened properties of the geopolymers were studied and correlated with the structure and microstructure of the geopolymers. The properties of the seeded geopolymers are explained in terms of the possible interactions between the seeds surface sites and the forming geopolymer matrix. The seeds had no effects during the first hours of reaction, but they affected the properties of the geopolymer at longer time. The zeolites seeds showed the most pronounced effects on the structural reorganization and compressive strength of the geopolymers.
KEYWORDS: geopolymer, seeding, nanoparticles, zeolites, microstructure, mechanical properties.

INTRODUCTION

Geopolymerization is a technology that converts aluminosilicates to binders which show performance advantages in respect to traditional binders. The advantages are related in particular to better resistance to acid and sulfate attack, thermal stability, possibility of heavy metals encapsulation, durability, etc. Geopolymers are attracting increasing attention due to two additional factors: the pressure to lower the CO_2 emissions during processing of cementitious binders (such as Portland cement) and to recycle industrial waste otherwise intended for disposal in the landfill. Moreover some ancient concretes resemble geopolymeric materials and thus they are historical prove of the outstanding durability of these new materials [1].

From a chemical point of view geopolymers consist of condensed silica and alumina tetrahedra with alkali cations balancing the negative charge on aluminum [2,3]. The empirical formula of geopolymer resembles the zeolite chemical formula and it is expressed as: $M_n((SiO_2)_z\,(AlO_2))_n \cdot wH_2O$ where "M" is a balance alkali metal cation such as K^+, Na^+ or Ca^{2+}; "n" is the degree of polycondensation; "z" usually varies between 1 and 3 and "w" is the water content.

In principle all materials containing silica and alumina are suitable for geopolymerization[4]. In practice the silica and alumina must be sufficiently reactive for the geopolymerization reaction to be appreciable; the more reactive the alumina and silica, the more and faster they will be converted into geopolymer. For this reason chemical analysis of the raw materials is not enough and X-ray diffraction techniques should be used to determine the amount of amorphous and thus reactive phase in the material. Curing temperature above room temperature can improve the reaction kinetics since the reactivity of material toward activating solution increases with the curing temperature [5,6]. The suitable precursor materials for geopolymerization can be roughly divided into two classes: 1) Thermally

activated alumino silicates, especially but not exclusively clay minerals containing kaolin[7] ; 2) Industrial by-products in which the alumina and silica bearing wastes have being subjected to high temperature and then abruptly cooled to room temperature. The two most used industrial by products in geopolymer applications are slag and fly ash. The precursors are mixed with an activating solution which can be an alkali metal hydroxide solution or a mixture of a soluble silicate with the former.
Metakaoline $Al_2Si_2O_7$ is an amorphous alumino-silicate obtained by calcination of kaolinite $Al_2Si_2O_5(OH)_2$. The high percentage of alumina, homogeneous distribution of silica and alumina and low percentage of contaminants in metakaoline make it a good raw material to model the geopolymerization and it has been extensively used in research for this purpose.
The first conceptual mechanism for geopolymerization was elaborated in 1959 by Glukhovsky. He attempted to explain the reactions involved in the formation of "soil cement", a new binder he was developing based on industrial wastes rich in aluminosilicate and alkalis. Glukhovsky's model described the geopolymerization reactions in three steps: 1) destruction - coagulation; 2) coagulation - condensation; 3) condensation - crystallization. The stages are so defined because they cannot be isolated since dissolution reactions proceed simultaneously with the gel formation and the polycondensation setting reactions. Starting from this concept, more sophisticated models have been proposed and used to simulate the heat flow in isothermal calorimetry experiments studying the geopolymer reactions[8] and to bring new insight on the parameters that govern the setting rate of geopolymers.
For metakaoline based geopolymer a maximum in compressive strength is observed to occur for sample with a Si/Al=2 [9,10], which can be defined as polysiloxo-sialate according to Davidovids terminology[1]. Since metakaoline has a ratio Si/Al=1, usually extra silica is provided in the form of soluble silicate. For sample with Si/Al <1.4 the microstructure is characterized by clustered dense particulate with large interconnected pores, while for Si/Al>1.65 the microstructure is homogeneous and the porosity is distributed to smaller pores non visible by SEM [10]. However the soluble silicates modify the reaction pathway and reaction products: according to Zhang [11] sodium silicate has a templating role. In fact geopolymers synthetized using sodium hydroxide contain crystalline zeolites, whereas the addition of even small amount of soluble silicate suppresses the zeolite crystallization [11,12]. For the relatively simple system including metakaoline and sodium hydroxide different zeolites have been detected[11,13-15]: zeolite A, zeolite X, zeolite Y (Na faujasite), hydrosodalite, sodalite and others.
In the last decade a number of research studies have focused on seeding of geopolymers [16-19]. Rees [16] used Al_2O_3 particles to study seeding of fly ash activated with 6M sodium hydroxide. He argued that silicate monomers quickly complex the liberated Al in solution, reducing the surface passivation and allowing more rapid fly ash dissolution. Nuclei induced by seeding would also remove dissolved components from the solution and yield enhanced ash dissolution.
Hajimohammadi [17] used Al_2O_3, ZnO and ZrO_2 particles to seed geopolymer made of geothermal silica and sodium aluminate. He argued the seeds: (i) induce nucleation of geopolymer gel in the initially Al-rich reaction mixture; (ii) prevent dissolved alumina from sorbing onto the surface of silica particles impeding their dissolution; and (iii) enhance the dissolution of silica.
Riahi [18] used rice husk ash, fly ash, sodium hydroxide and waterglass system cured at different temperatures to study the effects of amorphous nanosilica and nano α-alumina on the compressive strength of the material. Alumina showed no increase on strength, while the use of silica increased the mechanical strength by acting as heterogeneous nuclei and accelerator and promoter of the reaction.
Phair [19] studied the effects of ZrO_2 seeds on metakaoline and fly ash geopolymers activated with soluble silicates type-N, sodium hydroxide and sodium aluminate $NaAlO_2$. He concluded that zirconia is unable to act as a nucleation germ or template for crystalline polysialate (zeolite) formation.

The nucleation scenario is quite complex since silica and alumina are also used as reagents materials for geopolymerization, so some consider them as reagents and others as seeds. In the first scenario the particles will change the chemical equilibrium of the system, while in the second case they will only act as heterogeneous sites for the growth and crystallization of geopolymer.

In this preliminary study we investigate the seeding effect of silica and alumina particles on metakaoline-sodium hydroxide based geopolymer in terms of setting time, compressive strength and durability. The microstructure of the fractured surfaces was investigated using SEM and the structure studied by XRD and IR spectroscopy. Compressive strength, durability, microstructure and structure of the resulting geopolymers were compared with those of geopolymers seeded with pure synthetic zeolite, namely Mordenite and sodium and hydrogen Faujasite. These zeolites were selected since they have different Si/Al ratio and different site acidity. Surface activation of the zeolite was achieved by high intensity ball milling to study its effect on the abovementioned geopolymer properties.

EXPERIMENTAL

Materials

Metakaoline was purchased under the brand name of Powerpozz from Advanced Cement Technologies [Blaine, WA, United States]. Powder X-ray diffraction revealed the sample is predominately amorphous with traces of silica SiO_2, anatase TiO_2 and halloysite $Al_2Si_2O_5(OH)_4$. The chemical composition determined with SEM equipped with energy-dispersive X-ray spectroscopy (EDX) was 53.3% SiO_2, 43.5% Al_2O_3, 1.9% Fe_2O_3, and 1.3% TiO_2.

Sodium hydroxide solutions were prepared minimum 24 hours prior to use dissolving sodium hydroxide NaOH pellets (Alfa Aesar, 98%).

Different particles were used in order to study their effects on geopolymerization. 5 MIN-U-SIL Fine Ground Silica SiO_2 was purchased from U.S. Silica [Chicago, IL, Unites States], α-Al_2O_3 Nano Powder from Inframat Advanced Materials [Manchester, CT, Unites States], 99,99%, 100 nm average particle size distribution, was adopted as alumina nanoparticles seeds.

Three different zeolites were purchased from Alfa Aesar, Mordenite (Si/Al=6.5), sodium Faujasite type Y (Si/Al=2.5) and hydrogen Faujasite type Y (Si/Al=2.5). In order to modify the surface sites the zeolites, underwent high intensity ball milling using a "Spex 8000 mixer mill". The grinding media consisted of 3 mm zirconia balls, a ratio ball/powder of 50 and a milling time of 15 minutes were adopted since longer milling time resulted in amorphization of the material.

Preparation and curing of geopolymers

In the seeded geopolymer the seeds were added directly to the metakaoline powder and gently mixed with a spatula for 30 seconds. Then sodium hydroxide solution was added to the powders and mixed with an impeller mixer at approximately 2700 rpm for 3 minutes. The impeller was a stainless steel cross blade (18 mm x 30 mm). The slurries were cast into cylindrical polymeric molds with height and diameter of ≈3 cm. The cast specimens were placed on a vibrating table for 2 minutes to remove the entrapped air. After 15 minutes from the initial mixing the molds were placed in a water bath (Lab-Line Instruments 1500W) at a temperature of 45±3°C. The samples were unsealed and not in contact with water.

The general formula for this geopolymer is $Na_2O \cdot 2(SiO_2) \cdot (Al_2O_3) \cdot 10.12(H_2O)$, which is a polysialate according to Davidovids nomenclature [1]. In order to obtain this composition 11 M NaOH solution was added to the metakaoline powder with a ratio of solution/metakaoline equal to 1.1. Thus the total molar ratios of the geopolymer were: Na_2O/SiO_2 =0.5; Na_2O/Al_2O_3 = 1; SiO_2/Al_2O_3 = 2 ; H_2O/Na_2O =10.1. This formulation was adopted in order to avoid the templating effect of the soluble silicate[11] and focus on the seeding effects of oxides particles and zeolites. For the seeded geopolymers a 2% by weight of seeds (calculated vs the metakaoline content) was used. In Table 1 the codes for the different seeded geopolymers are shown. The chemical composition and curing conditions resembled the ones used by Zhan[6] who, using isothermal calorimetry, detected three exothermic peaks associated with the immediate dissolution of metakaoline, the condensation of oligomers and the structural reorganization of the geopolymers which occurred between 36 and 84 hours.

Table 1 Codes for the geopolymers containing different nanoparticles. MOR=Mordenite, Fau= Faujasite, m.=milled

sample code #	1	2	3	4	5	6	7	8	9
seed type	none	SiO_2	Al_2O_3	MOR	m.MOR	Fau-Na	m.Fau-Na	Fau-H	m.Fau-H

Test and analysis methods

Time of setting and unconfined compressive strength tests were performed on every formulation. The fractured surfaces were analyzed using SEM (scanning electron microscope). Portions of the fractured samples were then pulverized using agate mortar and pestle and analyzed using attenuated total reflectance Fourier transform infrared (ATR-FTIR) spectroscopy and powder X-ray diffraction (XRD). Bulk density, moisture content and durability of the intact samples were also determined.

Since there is not ad-hoc test for geopolymer, the setting times of the geopolymers were measured using a Vicat test (ASTM C 191 "Standard method for time of setting of hydraulic cement by Vicat needle").

The UCS (unconfined compressive strength) was tested using a INSTRON 3369 test machine equipped with a 50kN load cell and using a crosshead speed of 0.5mm/min. A total of six specimens were tested to evaluate the 3 days and 7 days compressive strength for each formulation. ATR-FTIR analyses were performed using a Perkin Elmer spectrum 100 - FTIR machine with a Tl-Br-I plate (diamond coated). The resolution adopted was 2 cm^{-1} and the spectra were acquired from 4000 cm^{-1} to 250 cm^{-1} averaging 25 scans.

X-ray powder diffractograms were obtained using a Rigaku Model MultiFlex with Cu-K_α (λ=1.5418Å) in Bragg Brentano theta-theta geometry with 1 cm receiving slit, 1° soller slit and 0.3° divergence slit. The scans were set to run at 40 kV and 30 mA with the step size ranging from 0.5 to 2 degree/min depending on the tested samples. XRD and ATR-FTIR analyses were performed on each material at three different aging times: 3 hours, 3 days and 7 days. The geopolymer samples were kept in a sealed bag to avoid water evaporation, crushed using mortar and pestle and immediately analyzed without drying. This procedure was adopted in order to study the geopolymerization products in the bulk material without alteration due to water removal.

The microstructures of the raw materials and of the fractured surfaces of the geopolymers at 3 days and 7 days were analyzed using a Hitachi Model S570 SEM with a tungsten filament and equipped with EDX. The materials were examined in high vacuum mode and the samples were gold-coated for 30 second using a sputtering machine in order to make them conductive and minimize charging effects.

Durability test were performed in accordance with Lancelotti[14]: undamaged samples were soaked for 24 hours in distilled water using solid/liquid ratio of 10 and the values of pH and conductivity were used to indirectly evaluate ions release and the stability of the geopolymeric matrix.

RESULTS AND DISCUSSION

The effects of high intensity ball milling on zeolite seeds

The effects were studied through FTIR and XRD. All the zeolite specimens after milling had broader peaks, both in the FTIR and in XRD. The peak broadening in XRD is due to a decrease in crystal sizes and due to the introduction of strains in the crystals. Figure 1 shows the ratios of full width half maximum (FWHM) of ball milled zeolite over as-received zeolite for the XRD diffractograms of mordenite. An increase of the FWHM was also observable for the FTIR peak of Mordenite, Figure 2, which indicates a larger distribution of bond lengths and angles probably due to the strain induced by ball milling. This trend was common for all the zeolites used in this work.

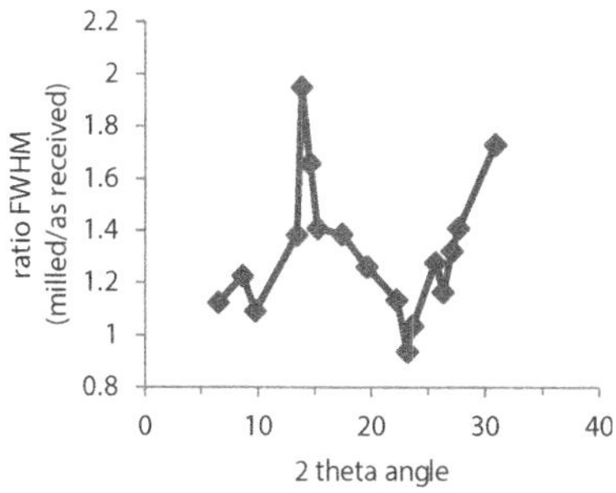

Figure 1 FWHM ratio of milled mordenite on as-received mordenite measured from the XRD diffractograms.

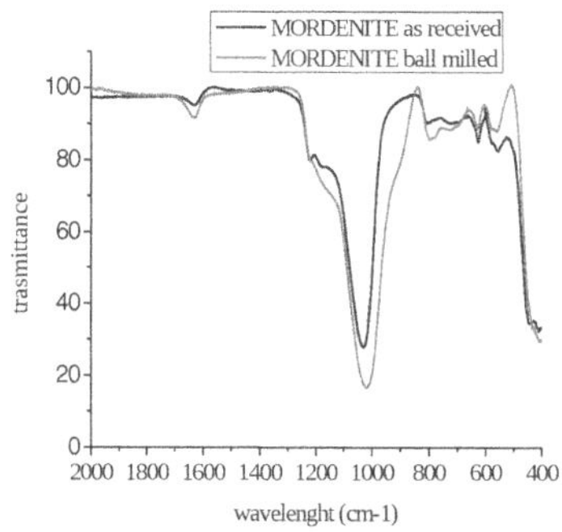

Figure 2 ATR-FTIR spectra for the as-received and ball milled mordenite.

Size strain analyses were performed on diffractograms acquired from 3° to 120° 2θ and the instrumental peak broadening was subtracted using the software X'Pert Highscore Plus. The peaks were fitted using pseudo-Voigts and the sample broadenings were analyzed using the Caglioti formula. The crystal sizes and strains of the as received and ball milled sodium Faujasite type Y were respectively 120 nm and 0.05%, 300nm and 0.14%. The results for the hydrogen Faujasite type Y were 150nm and 0.18%, 490nm and 0.23%. The Mordenite samples results were 2400 nm and 0.21%, 110 nm and 0.19%. The low intensities of the peaks at 2θ > 50° made it difficult for the algorithm to discern the contribution of size and strain: the increase of crystal size for the ball milled Faujasites samples is a consequence of it.

Setting time

The Vicat needle penetration values into the setting metakaoline-based geopolymers were more difficult to measure compared with Ordinary Portland Cement (OPC) pastes due to their highly thixotropic behavior. In particular the final setting time values were quite arbitrary because the needle left a small mark on the surface even when the geopolymer appeared to have set. This was possibly due to a lower hardness of the geopolymer as compared with OPC. The seeds had no evident effect on the setting time, which had the value of 75 minutes ± 15 minutes for all the samples.

Compressive strength, density and durability

Table 2 presents the results for unconfined compressive strength (UCS), bulk density and durability. The compressive strength is a highly variable quantity (has a large standard deviation) since it depends on density of defects, their size and distribution.

Table 2 Values of density and unconfined compressive strength at 3 days and 7 days for the 9 different samples. The values of pH and conductivity were measured on samples cured for 7 days, after 24 hours of soaking in distilled water. In brackets the standard deviations.

Sample Code #	Density [g/cc]		UCS [MPa]		pH	conductivity [milli MHO]
	3d	7d	3d	7d	7d	7d
1	1.62(2)	1.59(4)	4.7(1.3)	5.8(1.2)	12.44	15.5
2	1.53(1)	1.59(4)	8.2(0.5)	6(0.2)	12.42	13
3	1.64(3)	1.58(2)	4(3)	6(1)	12.35	11
4	1.67(2)	1.66(3)	6.3(3)	9(2)	12.35	10
5	1.56(1)	1.64(4)	8(2)	11(2)	12.31	9.5
6	1.63(1)	1.64(2)	4.5(1.5)	10(2)	12.09	3.5
7	1.56(2)	1.60(2)	5.2(2)	6.1(1.3)	12.41	14
8	1.64(1)	1.66(1)	5.3(1.3)	10(3)	12.3	8.5
9	1.58(9)	1.56(2)	5.2(1.3)	8.3(7)	12.39	11

The average moisture content of the geopolymers was 27±3%. Considering their average density of about 1.6 g/cm^3, the dry bulk density was 1.17 g/cm^3 in accordance with the values obtained by Liew [15]; thus these geopolymers could possibly be used as light-weight materials. There was no correlation between the values of density and compressive strength, which means that the difference in compressive strength cannot be ascribed to the variation in porosity introduced during the handling of the paste.

The values of the compressive strength expressed as % of that for the non-seeded geopolymer are displayed in Figure 3. The SiO_2 and Al_2O_3 seeds (samples 2 and 3) did not show significant improvements, while the zeolites seeds have shown a significant (i.e. nearly double) increase in compressive strength. The samples seeded with mordenite zeolites had a higher compressive strength already at 3 days. The compressive strengths from 3 to 7 days for samples 1,2 and 3 did not improve, while in the samples seeded with zeolite there was a continuous increase in compressive strength which indicates an ongoing geopolymerization process not present in the non-seeded paste. The effect of milling on zeolite is more difficult to explain since the milling operation improved the compressive strength for Mordenite seeds and Faujasite-H seeds, but seems to have no effect on the Faujasite-Na. Further studies are required, in particular it's recommended to test a higher number of samples in order to have a better statistic. In fact the observed sample means are plus or minus roughly 2 standard deviations from the population means with a 95% confidence, which means that with the current data is hard to draw any definitive conclusions since most of the results overlap. Previous work[10] on geopolymers with similar composition and cured at 40°C found a compressive strength of approximately 15±3 MPa. This higher value is probably due to a more dense matrix with less entrained air obtained by adopting 15 minutes long vibrating operations. The seeded samples had better durability, especially sample 6 seeded with Faujasite-Na (Figure **4**); there was a good correlation between pH and conductivity values (Figure **4**).

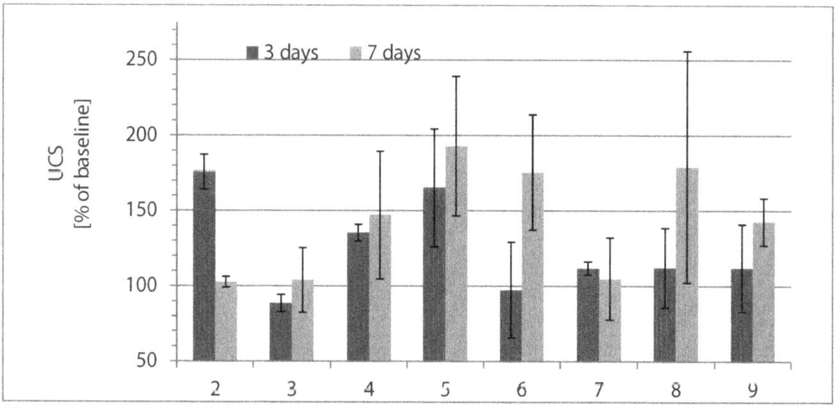

Figure 3 Percentages of the Unconfined Compressive Strength (UCS) of the seeded geopolymers compared with the non-seeded geopolymer at the same aging time.

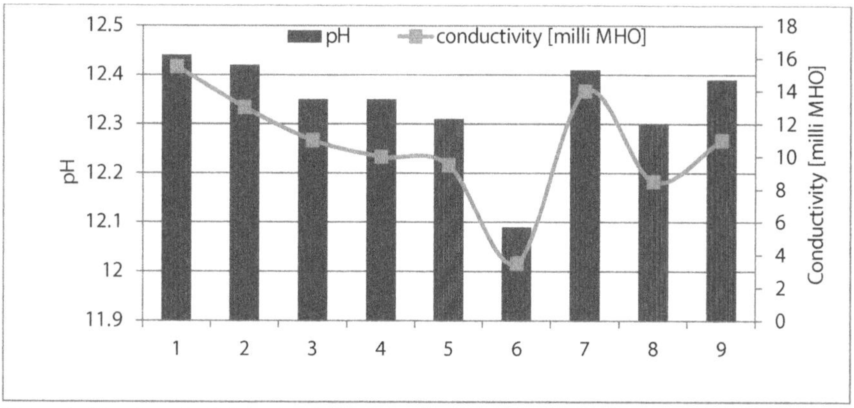

Figure 4 pH and conductivity values of the geopolymers soaked in distilled water for 24 hours.

Structure and microstructure of the geopolymers

Although all the geopolymers did set in 75 minutes, the XRD diffractograms of the seeded geopolymers (Figure 5) did not display any crystalline phases except for the small impurities already present in the metakaoline, including anatase, silica and halloysite. At 3 days crystalline peaks of zeolite type A - or Linde type A- and zeolite type X - which belong to the Faujasite family - appeared. The samples seeded with zeolites had even stronger zeolites peaks at 7 days which means that there was continuous nucleation and growth of zeolites in the geopolymer gel. This increase was also present for the sample seeded with alumina while there was not such an increase of crystalline zeolites for the sample 1 and 2. In Figure 6 shows the integrated intensities of the (200) reflection of zeolite A and

(111) reflection of zeolite X normalized over the intensity of the (011) reflection of the unreacted silica impurities.

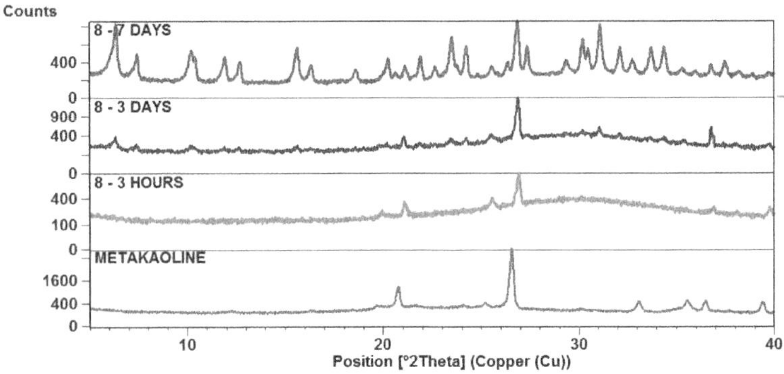

Figure 5 XRD diffractograms of metakaoline, milled-mordenite-seeded geopolymer at 3 hours, 3 days and 7 days.

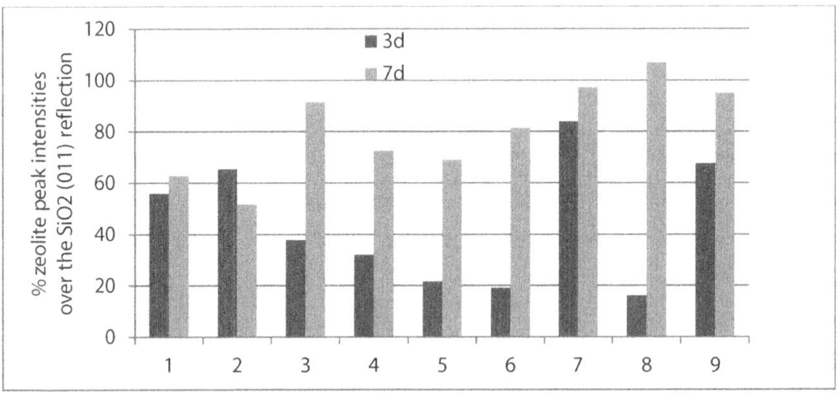

Figure 6 Zeolite content express in term of sums of integrated intensities of the zeolite crystalline phases over the (011) reflection of SiO_2 at 3 days and at 7 days.

This structural reorganization directly visible in the XRD and indirectly in the UCS values was seen in the FTIR spectra as well (Figure 7). The peaks associated with zeolite ring vibrations, which are located at low wavelengths [20], usually below 800 cm^{-1}, increased in intensity and became sharper at 3 days. For the samples 1, 2 and 7 there were no differences in the spectra between 3 and 7 days, while for all the other geopolymers the peaks became more intense and the FWHM decreased. The correlations between XRD, FTIR and UCS changes between 3 and 7 days indicated that the seeds influenced the structural reorganization of the geopolymers.

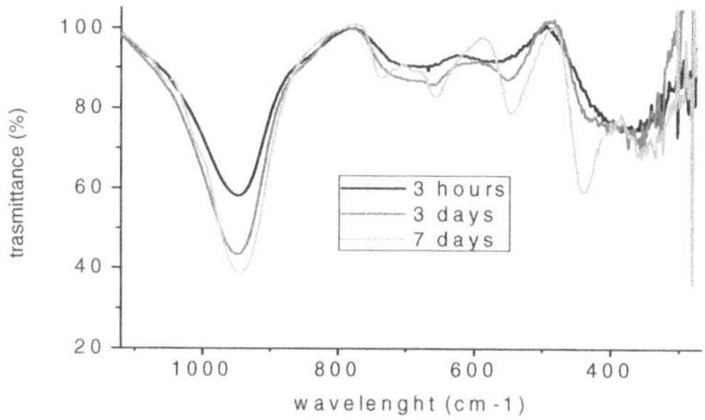

Figure 7 ATR-FTIR spectra for the sample 4, mordenite-seeded geopolymer, at 3 hours, 3 days and 7 days.

The microstructure of the geopolymers at 3 days (studied on the fractured surfaces) consisted of clustered particulates with large interconnected pores; at this time the materials were mainly composed of lumps though it was possible to detect some crystalline zeolites in the pores. At 7 days 5 μm cubic crystals were visible in most of the samples, together with smaller octahedral Faujasite-X crystals, approximately 0.5 to 1 μm large. Complex twinning was also present for all the samples, Figure 8.

Figure 8 SEM images at 3 days (left) and 7 day (right) for the alumina-seeded geopolymer, sample 3.

CONCLUSIONS

In this work the effects of silica, alumina and zeolites seeds on the geopolymerization of metakaoline based system were studied. The seeding had no evident effects on the short term properties of the geopolymers. In particular the different samples had a similar setting time equal to about 75 minutes, no crystalline phases and no definite zeolitic ring vibrations were detected after 3 hours. The setting reactions appear to correspond with the dissolution of metakaoline and the precipitation of amorphous geopolymer lumps with the characteristic FTIR band at 950 cm^{-1}. The structural reorganization that occurs between 3 and 7 days results in an increase in crystalline zeolite content detectable both in XRD diffractograms, ATR-FTIR spectra and SEM images and as an increase in compressive strength.

The seeds, with the exception of SiO_2 and milled-Faujasite-Na seeds, shown an enhancement in the structural reorganization compared with the non-seeded geopolymer. The geopolymers seeded with zeolites had higher compressive strength, crystallinity and durability. In particular the best result for compressive strength was obtained using milled-Mordenite which gives a 92% increase compared with the non-seeded geopolymer at 7 days.

REFERENCES

1. Davidovits J. *Geopolymer Chemistry and Applications, 3rd Ed.* Institut Géopolymère; 2011:632. Available at: http://books.google.com/books?id=mJBotwAACAAJ&pgis=1. Accessed February 17, 2014.

2. Davidovits J. Structural characterization of geopolymeric materials with X-ray diffractometry and MAS NMR spectroscopy. *Geopolymer.* 1988. Available at: http://scholar.google.com/scholar?hl=en&btnG=Search&q=intitle:Structural+characterization+of+geopolymeric+materials+with+X-ray+diffractometry+and+MAS+NMR+spectroscopy#0. Accessed February 24, 2014.

3. Davidovits J, Quentin S. GEOPOLYMERS Inorganic polymerie new materials. 1991;37:1633–1656.

4. Xu H, Van Deventer JSJ. The geopolymerisation of alumino-silicate minerals. *International Journal of Mineral Processing.* 2000;59(3):247–266. Available at: http://linkinghub.elsevier.com/retrieve/pii/S0301751699000745.

5. Ahmari S, Zhang L, Zhang J. Effects of activator type/concentration and curing temperature on alkali-activated binder based on copper mine tailings. *Journal of Materials Science.* 2012;47(16):5933–5945. Available at: http://link.springer.com/10.1007/s10853-012-6497-9. Accessed February 21, 2014.

6. Zhang Z, Wang H, Provis JL, et al. Quantitative kinetic and structural analysis of geopolymers. Part 1. The activation of metakaolin with sodium hydroxide. *Thermochimica Acta.* 2012;539:23–33. Available at: http://linkinghub.elsevier.com/retrieve/pii/S0040603112001475.

7. Ruiz-Santaquiteria C, Fernández-Jiménez a., Skibsted J, Palomo a. Clay reactivity: Production of alkali activated cements. *Applied Clay Science.* 2013;73:11–16. Available at: http://linkinghub.elsevier.com/retrieve/pii/S0169131712002700. Accessed February 20, 2014.

8. Duxson P, Provis JL, Lukey GC, van Deventer JSJ. The role of inorganic polymer technology in the development of "green concrete." *Cement and Concrete Research.* 2007;37(12):1590–1597. Available at: http://linkinghub.elsevier.com/retrieve/pii/S0008884607002001. Accessed February 20, 2014.

9. Rowles M, O'Connor B. Chemical optimisation of the compressive strength of aluminosilicate geopolymers synthesised by sodium silicate activation of metakaolinite. *Journal of Materials*

Chemistry. 2003;13(5):1161–1165. Available at: http://xlink.rsc.org/?DOI=b212629j. Accessed February 20, 2014.

10. Duxson P, Provis JL, Lukey GC, et al. Understanding the relationship between geopolymer composition, microstructure and mechanical properties. *Colloids and Surfaces A: Physicochemical and Engineering Aspects*. 2005;269(1-3):47–58. Available at: http://linkinghub.elsevier.com/retrieve/pii/S092777570500490G. Accessed January 21, 2015.

11. Zhang B, MacKenzie KJD, Brown IWM. Crystalline phase formation in metakaolinite geopolymers activated with NaOH and sodium silicate. *Journal of Materials Science*. 2009;44(17):4668–4676. Available at: http://link.springer.com/10.1007/s10853-009-3715-1. Accessed February 5, 2014.

12. Zhang Z, Provis JL, Wang H, Bullen F, Reid A. Quantitative kinetic and structural analysis of geopolymers. Part 2. Thermodynamics of sodium silicate activation of metakaolin. *Thermochimica Acta*. 2013;565:163–171. Available at: http://linkinghub.elsevier.com/retrieve/pii/S0040603113002141.

13. Heller-Kallai L, Lapides I. Reactions of kaolinites and metakaolinites with NaOH—comparison of different samples (Part 1). *Applied Clay Science*. 2007;35(1-2):99–107. Available at: http://linkinghub.elsevier.com/retrieve/pii/S0169131706001086. Accessed February 20, 2014.

14. Lancellotti I, Catauro M, Ponzoni C, Bollino F, Leonelli C. Inorganic polymers from alkali activation of metakaolin: Effect of setting and curing on structure. *Journal of Solid State Chemistry*. 2013;200:341–348. Available at: http://linkinghub.elsevier.com/retrieve/pii/S0022459613000789. Accessed February 20, 2014.

15. Liew YM, Kamarudin H, Mustafa Al Bakri a. M, et al. Optimization of solids-to-liquid and alkali activator ratios of calcined kaolin geopolymeric powder. *Construction and Building Materials*. 2012;37:440–451. Available at: http://linkinghub.elsevier.com/retrieve/pii/S0950061812005454. Accessed April 15, 2015.

16. Rees C a., Provis JL, Lukey GC, van Deventer JSJ. The mechanism of geopolymer gel formation investigated through seeded nucleation. *Colloids and Surfaces A: Physicochemical and Engineering Aspects*. 2008;318(1-3):97–105. Available at: http://linkinghub.elsevier.com/retrieve/pii/S0927775707010904. Accessed February 4, 2014.

17. Hajimohammadi A, Provis JL, van Deventer JSJ. Time-resolved and spatially-resolved infrared spectroscopic observation of seeded nucleation controlling geopolymer gel formation. *Journal of colloid and interface science*. 2011;357(2):384–92. Available at: http://www.ncbi.nlm.nih.gov/pubmed/21397245. Accessed February 20, 2014.

18. Riahi S, Nazari A. The effects of nanoparticles on early age compressive strength of ash-based geopolymers. *Ceramics International*. 2012;38(6):4467–4476. Available at: http://linkinghub.elsevier.com/retrieve/pii/S0272884212001241. Accessed February 28, 2014.

19. Phair JW, Deventer JSJV, Smith JD. Mechanism of Polysialation in the Incorporation of Zirconia into Fly Ash-Based Geopolymers. 2000:2925–2934.

20. Mozgawa W, Król M, Barczyk K, Science M. FT-IR studies of zeolites from different structural groups. 2011;(7):671–674.

NH_3BH_3 AND $NaBH_4$ ENCLOSED IN GEOPOLYMERS AND ZEOLITES

C. H. Rüscher, L. Schomborg, Z. Assi, J. C. Buhl
Institut für Mineralogie, Leibniz Universität Hannover, Germany

ABSTRACT

Ammonia-borane, NH_3BH_3 (= AB) and sodium tetrahydroborate ($NaBH_4$ = SB) can easily be dissolved in sodium aluminate and silicate solutions. Upon mixing they result in immediate gel formation and re-crystallization of AB and SB during drying. In case of AB-G IR-absorption spectra indicate that the geopolymer matrix (= G) consists of mainly sialate units (Si-O-Al-O). Heating experiments to 120, 150 and 300°C show the formation of polyaminoboranes (PAB) and polyiminoborane (PIB). The underlying reactions could be related to hydrogen release in two exothermic peaks observed at around 122°C and 160°C. In the course of further optimizations of the SB content in SB-G the Si/Al ratio of G could slightly increase with increasing the Si/Al ratio between 0.27 and 2.7. The water content of SB-G decreases linearly with the decreasing contribution of the geopolymer matrix. SB-SOD could be crystallized systematically following a hydrothermal soft chemical synthesis technique at 60°C, 80°C and 120°C. The effect of variations in particle and crystallite sizes could be correlated with effective hydrogen release and the water content in the matrix of the SB-SOD.

INTRODUCTION

Borohydrides have received some attention due to their remarkable gravimetric and volumetric hydrogen contents for hydrogen storage materials. However, there are significant problems for applications, e.g. unfavourable hydrogen releasing temperature, slow releasing rates, formation of borazine and other volatile gases, instabilities to moisture leading to uncontrollable hydrogen depletion, lack of reversibility. This has enforced the development of some scaffold scenarios for a better handling. E.g. for ammonia-borane (NH_3BH_3 = AB), it could be loaded into mesoporous graphitic carbon nitride (MGCN) through a solution-impregnation route [1], infiltrated into mesoporous silica SBA-15 [2] and silica hollow nanospheres [3]. ZIF-8, a zeolite type MOF (metal organic framework) could be used to immobilize Ni-nanoparticles obtaining a highly effective catalyst for hydrogen generation from hydrolysis of AB, as well [4]. Mesoporous silica, MCM-41, was investigated as an alternative means to stabilize or destabilize infiltrated ammonium borohydride, NH_4BH_4 (= ABH_2) [5].

The method of impregnation or an infiltration procedure could be traced back to Barrer's idea [6] for loading zeolites like A, X, Y with boron salts like $NaBH_4$ (= SB) or $Al(BH_4)_3$. However, this idea could never be realized. Another method to encapsulate the BH_4-anion into the sodalite cages was discovered by a direct hydrothermal synthesis technique [7-10]. This technique could be successfully used only for the sodalite (SOD) itself [10]. Hydrogen could thermally be released stepwise [10, 11] in the reaction of the BH_4-anion with water molecules provided by a matrix which glues the polycrystalline material together. However, for applications as hydrogen storage systems much higher densities of BH_4-anions are required. Presently alkaline solutions containing 30 wt% of SB at a maximum are used in fuel cell applications [12]. SB contains 10 wt% of hydrogen, which makes 1.2 l of hydrogen per gram of SB (density 1.07 g/cm^3) The release of hydrogen is doubled in the reaction with water, gaining 2.4 l of hydrogen per gram SB. It has been reported that a new composite material, SB enclosed in a geopolymer matrix (SB-G) could gain 2 l of hydrogen per gram of SB-G [13]. Thus highest hydrogen contents could be handled easily in geopolymer type materials and not in their crystalline relatives. It has also been reported that AB could be enclosed in a similar way in a geopolymer matrix, which protects the material from uncontrolled explosion during thermally induced hydrogen release [14]. AB contains about 21 wt% hydrogen and 2.13 l H_2 could theoretically be received from 1 g AB (3 mol H_2 per mol of NH_3BH_3). On heating up to 300°C there are two steps for hydrogen release through the formation of polyaminoboranes and polyiminoboranes [15]. The hydrogen release of the third mol of H_2 occurs at much higher temperature and reveal boron nitride (BN). According to Kim et al. [16] polyaminoboranes could be taken as a BN precursor phase with a yield of about 75% transformation in nitrogen steam at 1400°C.

It is the purpose of the present contribution to report some new details concerning the thermal hydrogen release of AB-G, the effect of changes in the nominal Si/Al ratio of the geopolymer in SB-G, and on the crystallization of SB-SOD for synthesis at 60, 80 and 120°C.

EXPERIMENTAL

XRD was carried out on as received NH_3BH_3 and on the synthesized samples directly after drying on a Bruker D8 powder diffractometer (CuK_α radiation, 2 theta range 5° to 80°, step width of 0.01° and measuring time of 12 sec/step). The obtained data were evaluated with STOE WinXPow software. SEM/EDX investigations were carried out on crushed and gold sputtered samples (JEOL SM-6390A). Thermogravimetric / Differential-Thermo Analyses (TGA/DTA, Setaram Setsys Evolution) were carried out under conditions as described in the text. FTIR spectra (Bruker Vertex 80v) were taken systematically using the KBr method (1 mg sample in 199 mg KBr). The hydrogen content of the chosen samples were determined by acid catalyzed hydrogen release (ACHR). The same apparatus and acid injection methods were used as described earlier in reference [10, 11]. Diluted acid (3% HCl) was added in an excess. The added volume of acid was varied between 5 and 8 ml and was subtracted afterwards from the shown volume at the gas syringe to register only the amount of released gas. The diluted acid was injected with an injection needle through the plug, so the apparatus remains gas tight. For every sample different masses in a wide range from 10 to 80 mg (Gel samples) and up to 450 mg for the sodalite samples were investigated in order to get more reliable results from a linear regression. With the linear regression, an amount of released hydrogen per 1 g sample was calculated. The released gas volume contained hydrogen, which was checked by the hydrogen-oxygen-reaction. Additionally the released gas was checked with a gas detector, showing that the released gas contained no CO_2 (detection limit < 1000 ppm).

RESULTS AND DISCUSSION

AB-G

AB-G samples could be prepared by dissolving AB in aluminate and silicate solution. In particular 245 mg AB (Fluka) were solved into a sodium-aluminate solution (250 mg $NaAlO_2$, Riedel de Haen , in 1.5 ml distilled H_2O) and the same amount into a sodium-silicate solution (310 mg Na_2SiO_3, Fluka , in 1.5 distilled H_2O). A complete (clear) solution was obtained by mild heating (max 60°C) within some minutes. Mixing of these solutions caused immediate gel precipitation. The gel was dried at 80°C in air for 3 h obtaining a white powder. Fig. 1 shows a typical SEM micrograph of AB-G at a magnification of 10000. A typical particle size of around 1 μm could be imaged. The typical particle size had a much finer distribution compared to that of the AB as-received sample. This could be related to the solution re-crystallization process of the crystals. For the AB sample only B and N were obtained by EDX analysis, whereas for AB-G very strong signals for Si, Al, and Na were seen, compared to some weak identifications of B and N. This indicated that AB was covered by the geopolymer.

Fig. 1 Typical SEM picture of AB-G

AB crystallized in space group I4mm, a = 5.244 A, b = 5.033 A. It has been observed by XRD that all main reflections for AB were also seen for AB-G. The average crystal size of AB in AB-G could be estimated for some selected peaks, and found to be in the range of 40-60 nm using the Scherrer equation. In addition to diffraction peaks of AB, the XRD pattern of the AB-G samples show a broad bump between 20 and 40° 2 theta centered at about 30°-32° 2 theta (d = 0.30 nm – 0.29 nm). This contribution is attributed to the geopolymer formed by the condensation of appropriate units by mixing the aluminate and silicate solution. Compared to the XRD pattern of a typical geopolymer of Si/Al ≈ 2 ratio, the peak position was shifted by 2-3° to higher 2 theta. This could be related to a lower Si/Al ratio of the geopolymer network for AB-G. A Si/Al ratio of about 1 was indicated by the characteristic peak of the geopolymer matrix at about 980 cm^{-1} (denoted by G in Fig. 2a), which was related to asymmetric stretching of the SiO_4 units within Q^4(4Al) groups. In comparison, for a geopolymer with a network of Si/Al ≈ 2 the asymmetric Si-O stretching is at 1016 cm^{-1} (KBr-method). Generally the peak position is sensitively dependent up variations in bondings of [SiO_4]-units bridged through the oxygen [16].

The peak characteristics of AB in the AB-G sample (Fig. 2a) can be identified by direct comparison with the spectrum of pure AB given in Fig. 2b where the peaks are assigned to N-B, B-H and N-H vibrations. The effect of heating to 120, 150 and 300°C as characterized by IR absorption of AB-G and AB samples are also given in Fig. 2a and b, respectively. Pure AB melts at about 115°C, where a strong evaporation of material occurs. Collecting the melt and pressing it quickly into the KBr-pellet yielded some indications of DADB (di-ammonitate of diborane, Fig. 2b) which could be described by

$$2\,(NH_3BH_3) \rightarrow [(NH_3)_2BH_2]BH_4 \qquad (1)$$

DADB was probably formed via hydride transfer without the loss of hydrogen [15]. Evaporation could occur with the formation of hydrogen via

$$NH_3BH_3 \rightarrow H_2 + H_2N\text{-}BH_{2(g)} \qquad (2)$$

Heating a bit longer at 120°C (10 min) then gave another result also shown in Fig. 2b. Now there was clear evidence that formation of polyaminoboranes (PAB) occurred according to:

$$NH_3BH_3 \rightarrow H_2 + 1/n\,(\text{-}H_2N\text{-}BH_2\text{-})_n \qquad (3)$$

Further heating steps at 150°C and 300°C showed partial and nearly complete transformation of PAB into polyiminoboranes (PIB), respectively:

$$1/n\,(\text{-}H_2N\text{-}BH_2\text{-})n \rightarrow H_2 + 1/n\,(\text{-}HN{=}BH\text{-})_n \qquad (4)$$

Similar products could also be obtained for AB-G samples following the same heating sequence of 120°C, 150°C, 300°C (Fig. 2a). Handling was, however, much easier as no evaporation loss occurred during melting. Moreover there were only very small indications of DADB as evidenced by the rather small peak at about 1123 cm^{-1} as marked in Fig. 2b.

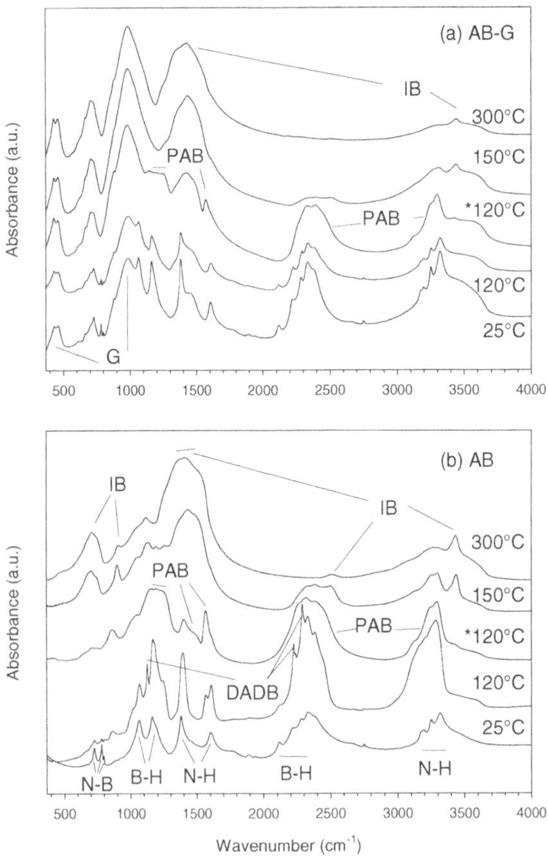

Fig. 2 IR absorption spectra of AB-G (a) and AB (b) at 25°C, and heated at 120°C, 150°C and 300°C. Assignment of peaks as denoted.

The effect of protection against evaporation and loss of material of AB embedded in the geopolymer was also rather helpful in TGA/DTA experiments. These experiments were carried out using He flowing gas (20 ml/min), and heating and cooling rates of 5°C/min and 10 min holding time at 300°C. The AB-G sample revealed a total mass in two main steps (Fig. 3). The differential thermogravimetric curves (dTG) showed peaks centered at about 122°C and 160°C. These peak positions coincided with two strong exothermic peaks which could also well be related to the hydrogen release reactions described in eq. 3 and 4. The heat flow showed some small but significant endothermic effects before a steep exothermic peak evolved. This effect could be related to the melting of AB in AB-G. As has been shown [14] the endothermic peak of melting of pure AB occurred at slightly lower temperature (115°C). The question arose if a mass loss due to the evaporation of any other gas as hydrogen from the AB-G could be ruled out. For comparison, TGA curves obtained by Frueh et al. [17] and Fang et al. [18] showed mass losses of about 20% between 110 and 130°C, followed by a second step which was finalized above about 180°C at a level of about 50-54%. Frueh et al. [17] could obtain by mass spectroscopy two maxima in hydrogen release, at 120 and at 155°C, which are consistent with the peaks as seen in the dTG-signals (Fig. 3a). Similar results were also obtained by Zhang et al. [3]. Since a total mass loss related to hydrogen could only be 19.6 wt% there must be a loss in N and B in the experiments in Refs. [17, 18]. The mass loss obtained for our AB-G sample amounted in total 11.8%. There was also some dehydration of the geopolymer, which could be seen to start from about 50°C. Thus we may count only the strong signals in dTG between 110 to 135 and 140 to 170°C as a loss due to evaporation of hydrogen of about 3 and 2 wt%, respectively. Assuming a complete reaction according to eq. 3 in the first step, this implies that the AB-G sample contained about 46 wt% AB. This value is in agreement with AB used as solid reactant together with $NaAlO_2$ and Na_2SiO_3.

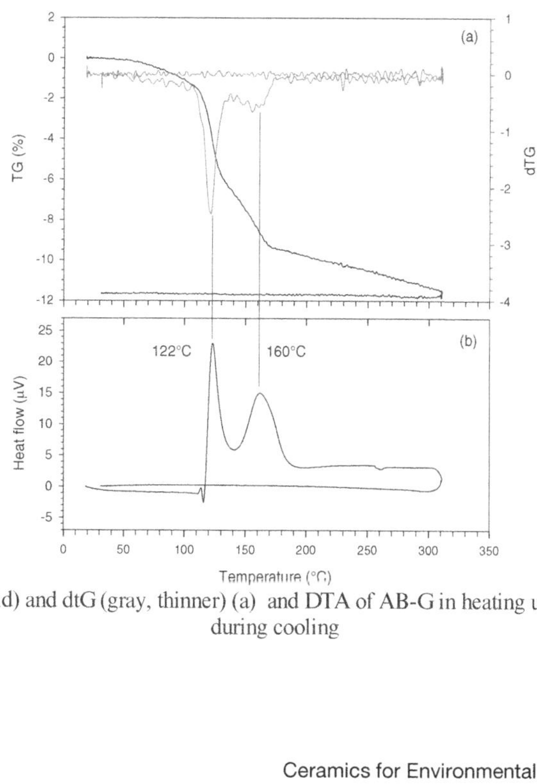

Fig. 3 TG (solid) and dtG (gray, thinner) (a) and DTA of AB-G in heating up to 300°C and during cooling

SB-G

Due to the excellent solution capabilities of SB in silicate and aluminate liquids it is very easy to handle. When mixed together the SB-containing aluminate and silicate solutions led to immediate gel-formation and secondary re-crystallization of SB. Systematic investigations concerning synthesis temperature, and optimum SB/matrix ratios (varying Si/Al ratio) for an optimum of hydrogen release as measured by ACHR, have been reported [13]. Some details concerning synthesis conditions and compositions of SB-G samples relevant to this study are collected in Table 1.

Table 1: Amounts of solids (SB, $NaAlO_2$, Na_2SiO_3), molar ratio of Si/Al, drying temperature (4h), SB to matrix (=$NaAlO_2$ + Na_2SiO_3) ratio, released hydrogen (in l) per g sample and percentage of effective enclosed SB (back calculated using the H_2 released with respect to SB given in the synthesis, for pure SB the experimental value of 2.24 l/g was used) for samples S1-S19.

Sample	SB (mg)	$NaAlO_2$ (mg)	Na_2SiO_3 (mg)	Molar Si/Al	T (°C)	SB/matrix	Released H_2 (l/g)	SB effective enclosed (%)
S1	1700	110	450	2.69	85	3	1.82	95.2
S2	560	400	160	0.27	110	1	1.06	83.5
S3	600	400	200	0.34	110	1	1.07	83.7
S4	500	300	200	0.45	110	1	1.12	88.3
S5	600	300	300	0.67	110	1	1.11	87.4
S6	500	200	300	1.01	110	1	1.10	86.6
S7	400	100	300	2.01	110	1	1.19	93.2
S8	750	150	600	2.69	110	1	1.21	95.2
S9	500	290	210	0.5	80	1	1.10	86.6
S10	500	200	300	1.0	80	1	1.19	93.2
S11	500	150	350	1.5	80	1	1.18	92.7
S12	500	120	380	2.0	80	1	1.26	99.2
S13	200	250	310	0.83	110	0.36	0.41	61.3
S14	500	250	310	0.83	110	0.89	0.93	77.7
S15	800	250	310	0.83	110	1.43	1.23	81.6
S16	1200	250	310	0.83	110	2.14	1.50	86.4
S17	1700	250	310	0.83	110	3.04	1.63	85.4
S18	500	250	310	0.83	85	0.89	1.08	90.1
S19	800	250	310	0.83	85	1.43	1.45	96.7

According to XRD the typical crystal sizes of SB are about 40 nm in average. The whole SB content appeared to be covered by the geopolymer matrix as shown in SEM micrograph of Fig. 4. EDX analyses indicated the presence of Si, Al and Na at any measurement spot. Optimization in the synthesis conditions with respect to the highest hydrogen release (182 ml H_2 per 100 mg sample = 95.2% effectively enclosed SB, sample S1 in Table 1) was obtained for composition SB/matrix ratio = 3.0 and drying temperature of 85°C. The IR absorption spectrum of this sample is shown in Fig. 5 (denoted as Si/Al // H_2 = 2.70* // 182) in the range between 700 and 1600 cm^{-1}. This spectrum is compared to the spectra of series of samples SB-G of SB/matrix ratios of 1 and Si/Al ratio from 0.27 to 2.7 (dried at 110°C, samples S2-S8, Table 1). The content of SB enclosed in the geopolymer is seen by comparing the characteristic peak for the $[BH]_4^-$

symmetric bending at 1125 cm^{-1} to the geopolymer peak at about 980 cm^{-1}, which is similar to the peak position also observed for AB-G. It can be seen that with increasing nominal Si/Al ratio, the peak gradually shifts to about 1010 cm^{-1} at Si/Al = 2.7. For samples S5, S6, S7 with nominal Si/Al = 0.67, 1.0, 2.0, there are indication that sodalite was formed, as seen by the triplicate peaks at 670, 700 and 760 cm^{-1}. This could also lead to a stronger additional contribution at 980 cm^{-1} in particular for samples with Si/Al = 1. The peaks at 840 and 880 cm^{-1} and between 1400 and 1500 cm^{-1} indicated the formation of carbonates due to the use of alkaline solutions under open conditions. A peak at about 915 cm^{-1} is observed for the sample with Si/Al = 0.67 which became gradually stronger with decreasing Si/Al ratio. According to Zhang et al. [19] this was assigned to Al-OH swing vibration of sixfold coordinated Al which could therefore not be relevant to the geopolymer.

Fig. 4 Typical SEM of SB-G (sample S1)

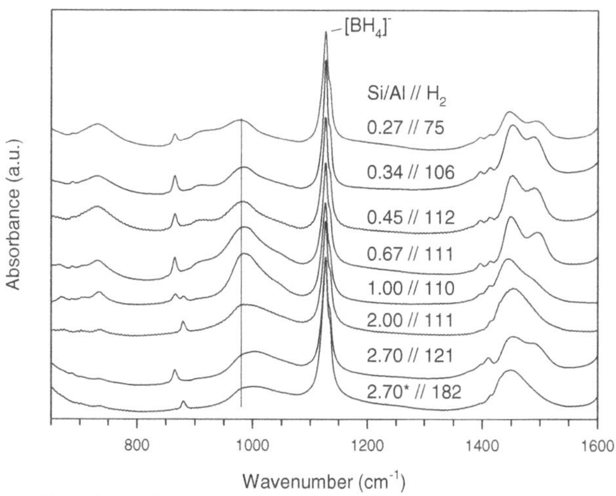

Fig. 5 IR absorption of samples SB-G S2-S8 and S1* (with Si/Al and released hydrogen in ml per 200 mg sample, for S2-S8 compare table 1).

A further investigation was carried out concerning the effective water (H_2O) content of the SB-G samples depending on Si/Al ratio, SB to matrix ratio and synthesis temperature. Typical TGA curves obtained for samples with SB/solid = 1 and Si/Al = 0.5, 1.0, 1.5, 2.0 synthesized at 80°C (samples S9-S12, Table 1) for heating up to 300°C are shown in Fig. 6. A mass loss of 3.5 to 4.5% was observed, indicating a slight increase with increasing Si/Al ratio. For another series of Si/Al = 0.83 synthesized at 110°C (samples S13-S16, Table 1) a linear decrease in mass loss related to dehydration with increasing $NaBH_4$/solid ratio was observed as shown in Fig. 7. This indicated that with a decreasing content of geopolymer there was a decreasing content of water suggesting that the water content was contained in the geopolymer matrix. Comparison with the results obtained for the 80°C synthesis series shows similar water contents enclosed in the geopolymer for SB/solid = 1.

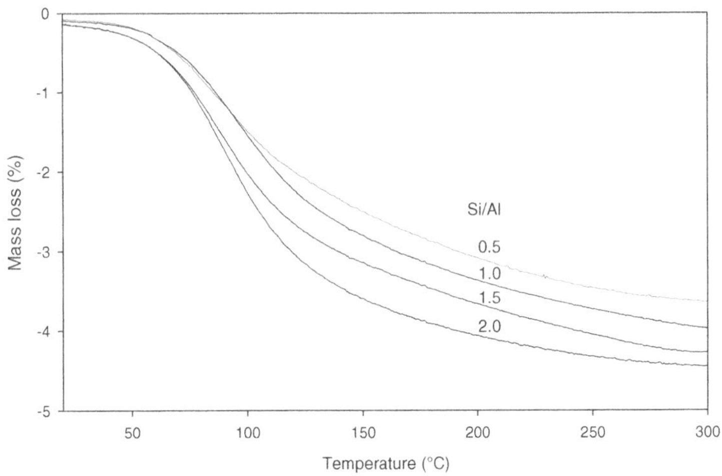

Fig. 6 TG curves of SB-G S9-S12 (table 1).

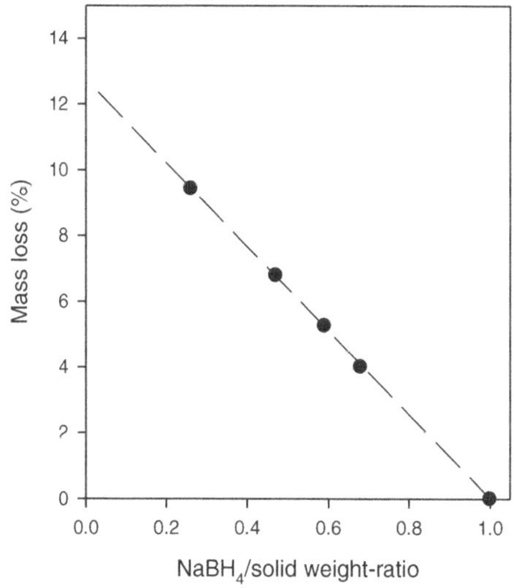

Fig. 7 Mass loss of SB-G samples S13-S16 (table 1).

SB-SOD

The enclosure of hydride in sodalites follows a completely different strategy compared to SB-G and AB-G composite materials. In SB-SOD the BH_4-anion is part of the SOD crystal structure, occupying the center in the sodalite cage. Synthesis procedures were followed as described for so called micro- and nano-crystalline SB-SOD [9]. The "microcrystalline synthesis" used a certain amount of SB, and 1 g of kaolinite (Fluka) in 10 ml of a freshly prepared 16 M sodium hydroxide (NaOH, Merck) which was added into a Teflon coated steel autoclave for reaction at 80°C and 120°C (24 h). In the "nanocrystalline synthesis procedure" a mixture of dissolved $NaAlO_2$ and Na_2SiO_3 was used in a wt-ratio of 0.4 to 0.6 in order to obtain the identical molar Si/Al ratio of 1 in the matrix. For the three synthesis batches a SB/matrix ratio of 0.8 was chosen to make sure, that the resulting samples consisted of highest contents of BH_4-sodalite, with as low as possible hydro-sodalite amounts. The syntheses were carried out under mild hydrothermal conditions in Teflon coated steel autoclaves for 6 h at 60°C. Sample materials were afterwards washed and dried.

Typical IR-absorption spectra of the three examples obtained at 60°C, 80°C and 120°C are shown in Fig. 8 in comparison to pure SB. The as-synthesized sample at 120°C revealed the four peaks of the BH_4-anion also seen for pure SB. The broad peak in the range between 3200 to 3600 cm^{-1} corresponds to O-H vibrations of water molecules, also showing some weak absorption at about 1630 cm^{-1}. These peaks became a bit more intense in the 80°C synthesized sample, and much more intense for so-called nanocrystalline SB-SOD. Peaks at 440, 450, the triplicate peaks between 600 and 800 cm^{-1} as well as the peak of highest intensity with maximum at 980 cm^{-1} are due to lattice vibrations of the sodalite framework. It could be ruled out that there was no significant contribution of BH_4-anions not structurally enclosed in the sodalite cage. First of all there were significant differences in intensity and position of the BH_4-characteristics in the pure salt at 1123, 2222, 2291, 2381 to be compared to BH_4^- in SOD at 1133, 2238, 2287, 2389 (in KBr, all in cm^{-1}). Secondly, XRD investigations show uniquely the presence of sodalite related peaks beside a broad peak at around 28° 2 theta. The later appeared to be rather weak for the 80 and 120°C synthesis but more significant for the 60°C batch. Evaluations of the sodalite peaks width using the Scherrer equation indicated average crystal sizes of about 80-100 nm for the "microcrystalline" batches, compared to about 50 nm for the samples from the "nanocrystalline synthesis route".

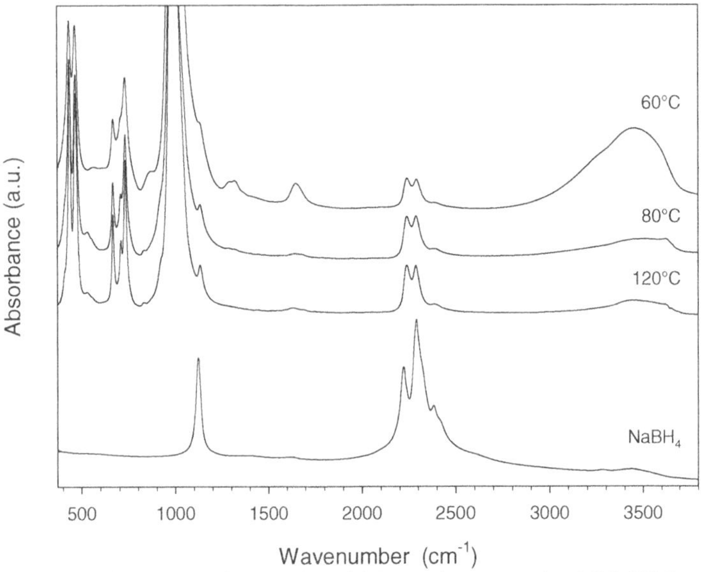

Fig. 8 IR absorption spectra of SB and SB-SOD samples prepared at 60°C, 80°C and 120°C (1mg in 100 mg KBr each).

Typical SEM micrographs of the three SB-SOD examples prepared at 60, 80 and 120°C and taken with the same magnification (10000) are compared in Fig. 9a-c. Most of the observed particles showed typical sodalite crystal morphology. We counted and measured their typical sizes in the microscope. The size of a particle was taken as an average of two perpendicular lines of their largest and smallest diameter. Their distribution function is shown in Fig. 9d. The 60°C batch yielded a maximum at around 290 nm (average of 224 particles). The batches prepared at 80°C and 120°C revealed maxima at around 525 (428 particles) and 564 nm (398 particles), respectively. It was the same trend as obtained by evaluating the XRD results using the Scherrer formula, however almost 5 times larger. This implied that the morphological unit with average size contains about 50 single crystals. For the 60°C batch the number may have been larger than 70. This also implies that the particles were compromise of crystalline and noncrystalline contents. The latter significantly increased with decreasing crystallite size. This could be seen by XRD, where the broad peak at around 28° 2 theta had a much higher intensity for the so called "nanocrystalline SB-SOD" compared to the "microcrystalline SB-SOD". It has been discussed earlier that this X-ray amorphous part in the sodalite, which might be seen like a geopolymer-type glue for sodalite crystals to form single morphological units, provided the water for the hydrogen release reactions of the BH_4-anions enclosed in the sodalite cage.

Fig. 9 SEM photographs of SB-SOD samples prepared at 60°C (upper left) , 80°C (upper right) and 120°C (lower left), and their particle size distribution (lower right)

Results on the total hydrogen content obtained in ACHR, the total mass loss and the reaction degree calculated from the loss in BH_4-anion cage fillings during thermal treatment are shown in Fig. 10. The total hydrogen content reads about 0.17 l and 0.18 l per 1 g SB-SOD prepared at 80°C and 120°C batches. Here the total mass loss obtained in TGA measurements in a heating run to 500°C counted about 2.2 and 1.8 wt% for the 80°C and 120°C batches. This loss was primary related to the effect of dehydration of the x-ray amorphous part of the crystal agglomerates. Additionally, a smaller part was related to the released hydrogen. This contribution could be measured by the decrease of the BH_4-related peaks in the IR absorption spectra. In the 60°C batch the total water content was much larger compared to the 80 and 120°C batches. A mass loss of about 8.5 wt% was obtained. As already mentioned, the amorphous fraction here is much bigger, so that the SB-SOD phase content was smaller. A total content of about 80 ml of hydrogen per g sample could be released by ACHR. On the other hand the degree of reaction, i.e. the decrease of BH_4-related peak intensity amounted about 85%.

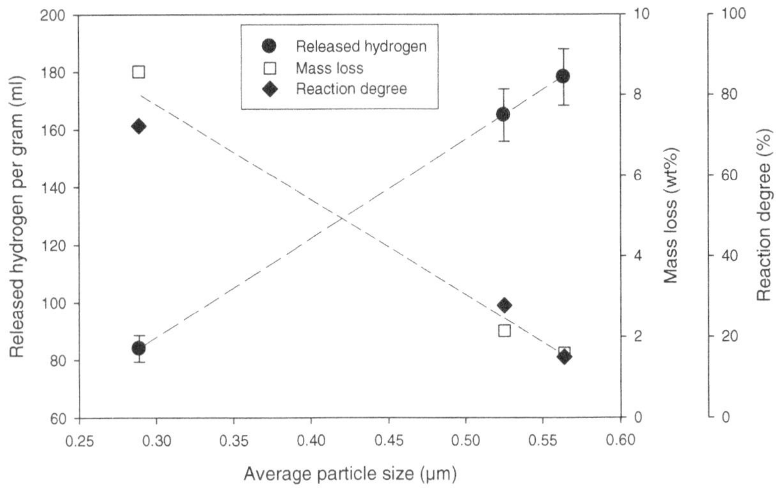

Fig. 10 Released hydrogen in ml per 1 g sample, mass loss (wt%) and degree of reaction of SB-SOD samples prepared at 60°C, 80°C and 120°C as a function of their average particle size.

CONCLUSIONS

NH_3BH_3-geopolymer (AB-G) and $NaBH_4$-geopolymer (SB-G) composite material could easily be prepared using AB and SB containing aluminate and silicate solutions. 1 g AB-G prepared using a AB/($NaAlO_2$ + Na_2SiO_3) wt-ratio of 0.89 and a Si/Al ratio of 0.8 revealed a total hydrogen release of 0.71 l in the ACHR (acid catalyzed hydrogen release). This could indicate that slightly less hydrogen could be released from AB in AB-G compared to AB, assuming a total of 47.2 wt% included in the geopolymer. For comparison, 1 g AB released 1.97 l by ACHR according to about 2.75 mol H_2 out of 3 mol available. Thermogravimetrically, two main processes related to exothermic reactions centered at 122°C and 160°C were observed, which could be related to hydrogen release with the formation of PAB and PIB, respectively.

An optimized SB-geopolymer (SB-G) could be prepared using a SB/($NaAlO_2$+Na_2SiO_3) wt-ratio of 3.04 and a molar Si/Al ratio of 2.7, dried at 85°C. 1 g of this sample released 1.82 l hydrogen in ACHR. This amounted to 82% obtained for pure SB in ACHR. The Si/Al ratio of the geopolymer matrix could be increased by increasing the Na_2SiO_3/$NaAlO_4$ ratio in the solution. The water content could exclusively be contained in the geopolymer matrix which did not affect the enclosed SB crystals during heating up to 300°C.

$NaBH_4$-SOD could be prepared hydrothermally in batches at 60, 80, and 120°C. Average crystal sizes were about 50 nm for the 60°C batch and 80 to 100 nm for the 80 and 120°C syntheses. The crystals formed larger units. The sizes of morphological units were measured and counted for some statistical evaluation. The 60°C sample contained the largest noncrystalline contribution in the morphological units depicting average size of about 290 nm. It contained the highest water content, 8.2 wt%, possessing the highest degree of thermally induced hydrogen release of about 82%, but the lowest total hydrogen content of 60 ml per g sample in ACHR. In contrast, the 120°C synthesized samples revealed the largest average crystal size (100 nm), agglomerated into the largest units (564 nm), exhibited the lowest water content (1.8 wt%) and lowest degree of reaction after heating to 500°C (15%) but the highest hydrogen contents in ACHR (180 ml/g sample) among the sodalites.

ACKNOWLEDGEMENTS

Some results obtained by ZA and LS were given prior to publication in her Ph. D. thesis work, supported by DAAD. Part of the work was supported by DFG under RU 764/6-1.

REFERENCES

[1] Z. Tang, X. Chen, H. Chen, L. Wu, X. Yu: Metal-free catalysis of ammonia-borane dehydrogenation/regeneration for a highly efficient and facilely recyclable hydrogen storage material. Angew. Chem. Int. Ed. 52, 5832-5835 (2013).

[2] A. Gutowska, L. Li, Y. Shin, C. M. Wang, X. S. Li, J. C. Linehan, R. S. Smith, B. D. Kay, B. Schmid, W. Shaw, M. Gutowski, T. Autrey: Nanoscaffold mediates hydrogen release and the reactivity of ammonia borane. Angew. Chem. Int. Ed. 44, 3378-3582 (2005).

[3] T. Zhang, X. Yang, S. Yang, D. Li, F. Cheng, Z. Tao, J. Chen: Silica hollow nanospheres as new nanoscaffold materials to enhance hydrogen releasing from ammonia borane. Phys. Chem. Chem. Phys. 13, 18592-18599 (2011).

[4] P.-Z. Li, K. Aranishi, Q Xu: Zif-8 immobilized nickel nanoparticles: highly effective catalysts for hydrogen generation from hydrolysis of ammonia borane. Chem. Commun. 48, 3173-3175 (2012).

[5] T. K. Nielsen, A. Karkamkar, M. Bowden, F. Besenbacher, T. R. Jensen, T. Autrey: Methods to stabilize and destabilize ammonium borohydride. Dalton Transactions 42, 680-687 (2013).

[6] R. M. Barrer: Hydrothermal chemistry of zeolites. London Academic Press, 384 p (1982).

[7] J. C. Buhl, T. Gesing, C. H. Rüscher: Synthesis, crystal structure and thermal stability of tetrahydroborate sodalite $Na_8[AlSiO_4]_6(BH_4)_2$. Microp. Mesopor. Mater. 80, 57-63 (2005).

[8] J. C. Buhl, T. Gesing, T. Höfs, C. H. Rüscher: Synthesis and crystal structure of gallosilicate- and aluminogermanate tetrahydroborate sodalites $Na_8[GaSiO_4]_6(BH_4)_2$.and $Na_8[AlGeO_4]_6(BH_4)_2$. J. Solid State Chem. 179, 3877-3882 (2006).

[9] J. C. Buhl, L: Schomborg, C. H. Rüscher: Tetrahydroborate sodalite nanocrystals: Low temperature synthesis and thermally controlled intra-cage reactions for hydrogen release of nano- and micro-crystals. Microp. Mesopor. Mater. 132, 210-218 (2010).

[10] J. C. Buhl, L. Schomborg, C. H. Rüscher: Enclosure of sodium tetrabydroborate ($NaBH_4$) in solidified aluminosilicate gels and microporous crystalline solids for fuel processing, in Hydrogen Storage, J. Liu ed., INTECH ISBN 978-953-51-0371-6, free online editons, Chapter 3, 49-90, (2012).

[11] C. H. Rüscher, L. Schomborg, A. Schulz, J. C. Buhl: Basic research on geopolymer gels for production of green binders and hydrogen storage. In Developments in strategic materials and computational design IV ed. by W. M. Kriven, J. Wang, Y. Zhou, A. L. Gyekenyesi, Ceramic engeneering and science proceedings, Vol. 34, 97-114 (2013).

[12] C. Ponce de Leon, F. C. Walsh, D. Pletcher, D. J. Browning, J. B. Lakeman, Review: Direct borohydride fuel cells. J. Power Sources 155, 172-181 (2006). Compare also Z. P. Li, B. H. Liu, K. Arai, K. Asaba, S. Suda. Evaluation of alkaline borohydride solutions as the fuel for fuel cell. J. Power Sources 126, 24-33 (2004).

[13] L. Schomborg, C. H. Rüscher, J. C. Buhl, F. Kiesel: $NaBH_4$ geopolymer composites. In Developments in strategic materials and computational design V. Ed. W. M. Kriven, D. Zhu, K II Moon, T. Hwang, C. Lewinsohn, Y. Zhou, Ceramic engeneering and science proceedings, Vol. 35, 3-14 (2014)

[14] L. Schomborg, C. H. Rüscher, J. C. Buhl, M. Wark: NH_3BH_4 geopolymer composite. In press. Ceramic engeneering and science proceedings, Vol. 35 (2015).

[15] C. W. Hamilton, R. T. Baker, A. Staubitz, I. Manners: B-N compounds for chemical hydrogen storage. Chem. Soc. Rev. 38, 279-293 (2009).
[16] D.-P. Kim, K.-T. Moon, J.-G. Kho, J. Economy, c. Gervais, F. Babonneau: Synthesis and characterization of Poly-(aminoborane) as a new boron nitride precursor. Polym. Adv. Technol. 10, 702-712 (1999).
[17] S. Frueh, R. Kellet, C. Mallery, T. Molter, W. S. Willis, C. King'ondu, S. L. Suib: Pyrolytic decomposition of ammonia borane to boron nitride. Inorg. Chem. 50, 783-792 (2011).
[18] Z. Fang, J. Luo, X. Kang, H. Xia, S. Wang, W. Wen, X. Zhou, P. Wang: Facile solid-phase synthesis of the diammoniate of diborane and its thermal decomposition behavior. Phys. Chem. Chem Phys. 13, 7508-7513 (2011).
[19] Y. Zhang, W. Sun, Z. Li: Infrared spectroscopy study of structural nature of geopolymeric products. J. Wuhan Univ. Technol. Sci. Ed. 23, 522-527 (2008).

Macroporous Ceramics for Environmental and Energy Applications

SILICON CARBIDE MEMBRANES FOR WATER FILTRATION APPLICATIONS

Melanie Kuhn[1], Abhaya Bakshi[1], Eric Sheridan[1], Fabiano Rodrigues[2], Adrien Vincent[2], Malte Moeller[3], Ronald Neufert[3]
[1]Northboro Research & Development Center, Saint-Gobain Innovative Materials, 9 Goddard Road, Northboro, MA 01532 USA
[2]Saint-Gobain, CREE, 550 Avenue Alphonse Jauffret, BP. 20224, 84300 Cavaillon, France
[3]Saint-Gobain IndustrieKeramik Rödental GmbH, 96466 Rödental, Germany

ABSTRACT

Porous silicon carbide (SiC) membranes are proposed for filtration applications for produced water in the oil and gas industry. Key advantages of SiC membranes include their chemical resistance, high temperature stability and low fouling behavior. Hence SiC-based filters can easily handle wastewater at elevated temperatures and are resistant to chemical cleaning. In addition, they have been shown to exhibit high flux compared to polymeric and other ceramic membrane materials. SiC cross-flow filters are manufactured by Saint-Gobain via a multi-step process where the SiC membrane is applied onto the channel walls of a porous SiC carrier structure. The membrane pore size is tailored through controlling various process parameters during manufacturing. SiC membranes with various pore sizes have been studied for the filtration of different types of produced water. For both synthetic produced water and produced water sourced from conventional and unconventional (fracking) oil wells, the SiC membranes were found to effectively remove suspended solids, bacteria and oil. Compared to current technologies used for the treatment of effluent produced water from oil and gas exploration, SiC membrane technology promises a smaller footprint, an overall more compact system and reduced usage of chemicals. These advantages also make SiC membranes attractive for other water treatment applications, such as in swimming pools.

INTRODUCTION

Water is a key element of our daily lives, and efficient treatment technologies are needed for cleaning and recycling water. Clean drinking water should be made available to every human being. In addition, increasing amounts of wastewater from industries, mining and oil well operations require treatment and/or recycling. Finally, water treatment even touches upon many other aspects such as providing clean, less skin-irritating and less chemically treated water in swimming pools. Technologies vary significantly depending on the type of water to be handled and can quickly become a multi-step, highly complex procedure, especially if very dirty, polluted wastewaters need to be cleaned.

With changing regulations and limited availability of fresh water, treatment of produced water from oil operations has gained increasing interest. Several treatment technologies are available, and references [1-3] provide a summary of many of these methods.

Membrane filtration for water treatment is based on the use of a porous media for the separation of compounds from the liquid stream. Depending on the pore size of the separating media, one can distinguish between microfiltration (0.1-3 μm), ultrafiltration (0.01-0.1 μm), nanofiltration and reverse osmosis (as small as 0.0001 μm).[2] Membranes can be made either from polymeric or ceramic materials. Ceramic membranes have the advantage of chemical resistance to a wide range of water types and cleaning materials. Also they can handle water at elevated temperatures where the use of polymeric materials would be limited.

Saint-Gobain manufactures silicon carbide (SiC) ceramic membrane filters for various microfiltration applications. These filters remove suspended solids, bacteria, oil and some cations.[4,5] Advantages of SiC membranes include chemical stability over the whole pH range, high temperature resistance as well as controlled pore size.

SiC ceramics can be densified by various methods such as pressureless sintering, nitride-bonding, oxide-bonding, reaction-bonding or recrystallization.[6,7] Recrystallized SiC is consolidated at high temperatures (> 2200 °C) through sublimation of fine SiC particles that condense at the larger SiC grain boundaries, resulting in a pure (>99%), strongly bonded and, most importantly, porous SiC body.[6-9] Recrystallized SiC exhibits high temperature stability and good thermal shock resistance; its high porosity structure makes it suitable for filtration applications, including filtration of gaseous (e.g. diesel particulate filters (DPF))[10] and liquid (e.g. water filtration) media.[9]
In the following sections, we review Saint-Gobain's SiC membrane technology and discuss different applications for water treatment using such membranes.

SAINT-GOBAIN SiC CERAMIC FILTERS

Manufacturing method

Saint-Gobain SiC filters and membranes are manufactured via a multi-step process wherein the SiC membranes are coated onto a SiC filter carrier structure. The filter carrier structure is manufactured by first preparing a paste using SiC powder in the desired grain size range and suitable additives, followed by extruding the paste and creating a filter structure with a certain number of channels and a certain channel design. The filter is sintered to create a homogenous microstructure through an evaporation-recrystallization process. Similarly, a separate SiC-based slurry with defined grain sizes is prepared which is then applied onto the channel walls of the filter carrier. A subsequent sintering step creates a homogeneous membrane microstructure with a controlled, tight pore size distribution.

Filter characteristics

Typically, the filter carrier exhibits an average pore size of ~25 μm while Saint-Gobain has the capability of manufacturing membranes with 250 - 1000 nm average pore size. Porosity is >40% for both membrane and carrier. Figure 1 shows cross-sectional views of both the carrier and membrane for a cross-flow SiC filter.

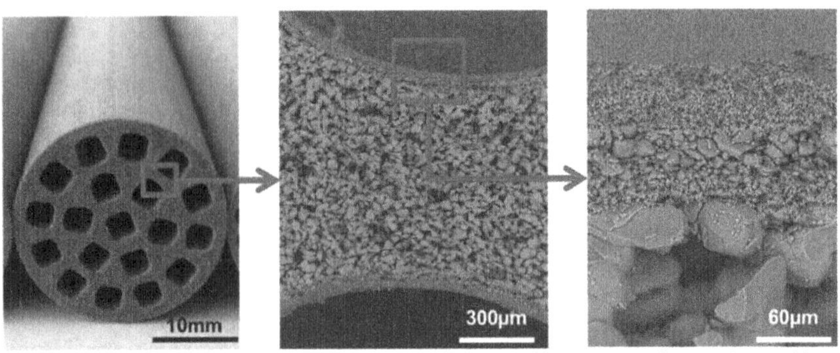

Figure 1: From left to right: Saint-Gobain cross-flow SiC filter (25 mm diameter); SiC filter carrier structure (coarse-grain) and SiC membrane (fine-grain); Close-up of carrier and membrane.

Cross-flow vs dead-end filters

Saint-Gobain currently manufactures two configurations of SiC filters – cross-flow and dead-end. The cross-flow configuration is commonly used for applications involving water with high solid loadings. The water enters the filter on one end, clean water or permeate (or filtrate) leaves the filter through the circumference while the remaining, unfiltered water, or retentate, exits at the opposite end and can be re-circulated back into the water stream (Figure 2).

Dead-end filters are suited for "clean" water applications, such as swimming pools or drinking water.[4] These honeycomb-structure filters are similar to diesel particulate filters which remove particles from diesel car exhaust streams. In the dead-end configuration, the channels are plugged either at the entrance or exit side; every other channel is plugged at the entrance side but is left open on the exit side. Feed water enters into such a channel, is forced through the membrane and then exits through the unplugged channels on the exit side (Figure 3).

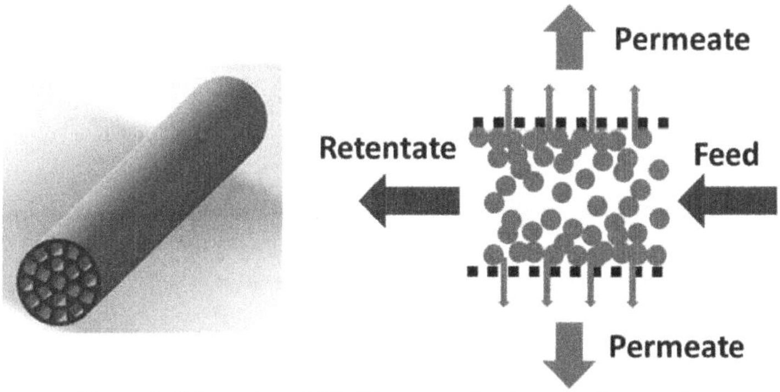

Figure 2: Schematics of SiC filter operating in cross-flow mode.

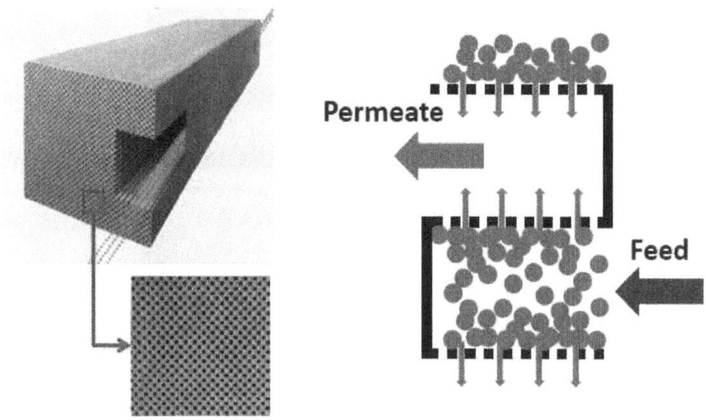

Figure 3: Schematics of dead-end SiC filter configuration.

FILTRATION TRIALS

Synthetic produced water

For in-house testing, a stable emulsion was prepared by mixing DI water, 4000 ppm NaCl, 100 ppm clay and 300 ppm mineral oil using a high shear speed mixer. The emulsion had a trimodal particle size distribution of 0.8, 3 and 20 μm. Cross-flow filters with 250 and

1000 nm membranes were tested. As can be seen in Figure 4, similar flux values were observed for both membrane sizes indicating the build-up of a filtration cake on top of the membrane surface. Backwashing at a 2 hour interval removed the filtration cake and restored the initial high flux before the filtration cake would build up again. The permeate or filtrate exhibited a much lower turbidity than the initial feed water (Figure 5).

In a comparative study, Saint-Gobain SiC membranes yielded a 25% higher permeability vs. oxide membranes based on TiO_2/ZrO_2.

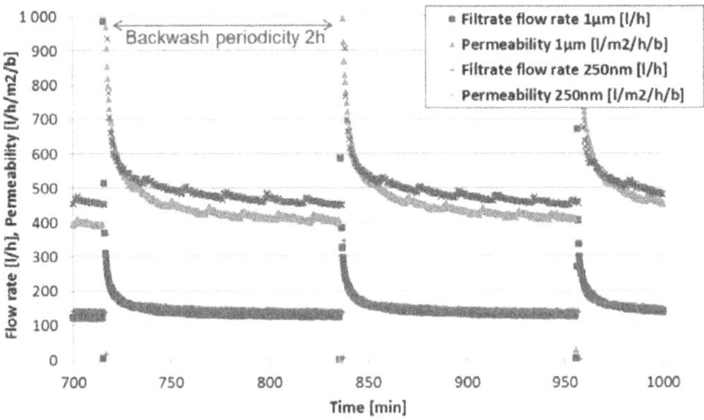

Figure 4: Flow rate and permeability as a function of time for filtration tests with synthetic produced water using 250 and 1000 nm Saint-Gobain SiC membranes.

Figure 5: Turbidity of synthetic produced water before and after filtration through a 250 nm membrane.

Produced water

Tests were performed using produced water from conventional oil wells in the Uintah Basin, UT, in collaboration with Environmental Clean Systems, LLC (ECS), a water management company based in Phoenix, AZ.[5] The produced water underwent some pre-treatment to remove H_2S and coagulate fine suspended solids. Both 250 and 1000 nm SiC membranes in the cross-flow configuration were evaluated. Figure 6 shows the flux for the 250 nm membrane as a function of time. Similar to the in-house testing using synthetic produced water (Figure 5), flux decreased over time due to the build-up of a filtration cake but could be restored by applying a backwash procedure. Backwashing had not been optimized during this test, which resulted in less than 100% recovery of the initial flux. Backwashing and CIP (Clean in Place) will need to be tailored for the type of water used and are an integral part of the overall design of the filtration equipment.

A stable flux of over 200 LMH was measured for a trans-membrane pressure (TMP) of ~1 bar. Care has to be taken in choosing the appropriate TMP as at high TMP, oil particles, which are deformable, can get squeezed into the membrane pores and can cause blocking of the membrane.[11]

Table I compares the water quality in terms of TOG (Total Oil and Grease), iron content, hardness and level of bacteria for the produced water, the produced water after pre-treatment and finally the pre-treated produced water after filtration through the Saint-Gobain SiC membrane. The permeate water ("after microfiltration") showed a clear reduction in both TOG and iron content. Using an additional pH treatment, hardness could also be significantly decreased. The level of bacteria was determined by using a serial dilution method for sulfate reducing bacteria (SRB) and acid producing bacteria (APB).[5] "4+" implies that all dilution vials contain bacteria while "0" represents full removal of both SRB and APB. While some bacteria were still left after passing the water through the SiC membrane, an additional pH pre-treatment resulted in the full removal of all bacteria in combination with microfiltration.

Although some pre-treatment is desirable to flocculate suspended solids and remove H_2S, the overall complexity and footprint of a pre-treatment + SiC membrane microfiltration system is significantly reduced when compared to a traditional water treatment plant where large settling tanks, oil-water separation, bag filters, etc. are needed.[5] In addition, through the efficient removal of bacteria through the SiC membrane, less biocide and chemicals are needed to post-treat the water and meet the required specifications for re-injecting the recycled water into the wells.

Figure 7 compares the turbidity of the produced water after pre-treatment and after microfiltration. The permeate is very clear, turbidity could not be measured.

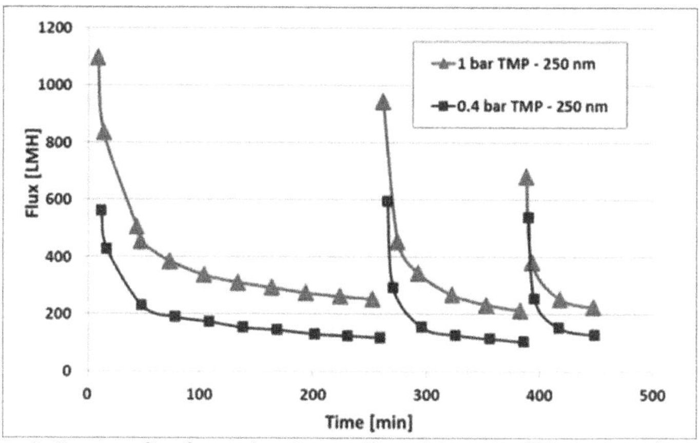

Figure 6: Permeate flux for produced water filtered through a 250 nm cross-flow SiC membrane at 0.4 and 1 bar TMP.

Table I: TOG (Total Oil and Grease), iron content, hardness and bacteria levels for produced water as-received, water after pre-treatment and after filtration through the SiC membrane. *Additional pH treatment.

	Produced water	After pre-treatment	After microfiltration
TOG [ppm]	Up to 150	86	12
Iron [ppm]	1-150 +	35	7.1
Hardness	200-20,000 +	475	35*
Bacteria	4+	4+	0*-2/3

Figure 7: Turbidity of feed water (produced water after some pre-treatment) and permeate after microfiltration through a 250 nm SiC membrane.

In addition to produced water from the Uintah Basin, produced water from several other oil well locations throughout the US, both conventional and fracking wells, was tested. Figure 8 shows the low turbidity of the filtered water.

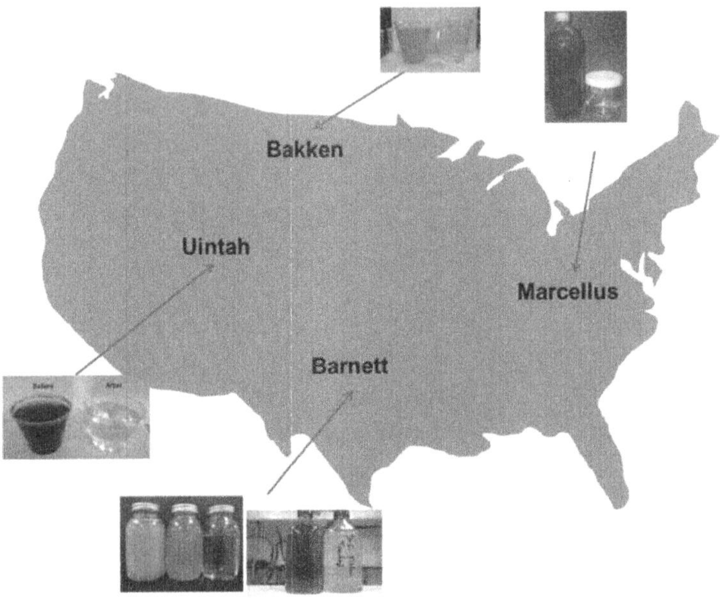

Figure 8: Produced water from oil wells across the US was passed through Saint-Gobain SiC cross-flow filters resulting in a significant reduction in turbidity.

Swimming pool water

Saint-Gobain has supplied and installed SiC dead-end filters in several swimming pools in Germany, with the first system having been online since 2012 (Figure 9). All systems have been working without any issues, and the water quality has been continuously meeting the requirements of German pool water quality regulations according to DIN 19643-1.[12] In addition to a smaller footprint compared to traditional sand filters, cleaning of SiC dead-end filters involves less water consumption and restores the filter to an "as-delivered" state after each cleaning. During operation, regular backwashing is performed to remove built-up filtration cakes and restore and maintain flux. Through the combination of SiC microfiltration, UV treatment and chlorine-electrolysis, high water quality can be achieved with a minimized "chlorine" pool-smell. The SiC filters provide a real barrier for bacteria and protozoae which reduces the required amount of chlorine. Further results on using Saint-Gobain SiC dead-end filters for swimming pool applications can be found in a previous paper.[4]

Figure 9: Saint-Gobain SiC dead-end filters in a swimming pool installation.

CONCLUSIONS

SiC membrane filters are an upcoming technology for microfiltration of water in various applications, ranging from clean water applications to treating hot, low/high pH produced water and industrial wastewater. Filtration systems using SiC membranes promise a smaller foot print compared to conventional water treatment systems. Due to their chemical resistance and robustness, they can be operated under harsh pH and temperature conditions, and they withstand any chemical cleaning procedure. Saint-Gobain SiC membrane filters have shown very positive results for use in swimming pools as well as for treating produced water from oil wells.

REFERENCES

[1] Arthur et al., "Technical summary of oil and gas produced water treatment technologies", ALL Consulting LCC, 2005.

[2] Igunnu and Chen, "Produced water treatment technologies", International Journal of Low-Carbon Technologies, vol. 9, pp. 157-177, 2014.

[3] Colorado School of Mines, "An integrated framework for treatment and management of produced water – Technical assessment of produced water treatment technologies", RPSEA Project 07122-12, 2009.

[4] Neufert et al., "Dead-End Silicon Carbide Micro-Filters for Liquid Filtration", Advances in Bioceramics and Porous Ceramics VI: Ceramic Engineering and Science Proceedings, vol. 34, pp. 115-126, 2013.

[5] Bakshi et al., "Treatment of produced water using silicon carbide membrane filters", Ceramic Engineering and Science Proceedings Volume 36, Issue 5, Advances in Bioceramics and Porous Ceramics VIII, Roger Narayan and Paolo Colombo, Editors; Jingyang Wang and Soshu Kirihara, Volume Editors; Copyright The American Ceramic Society, 2015.

[6] Lee and Rainforth, "Ceramic Microstructures - Property control by processing", Chapman & Hall, 590 p., 1994.

[7] Kriegesmann, "Sintering phenomena in silicon carbide", Powder Metallurgy International, vol. 18, pp. 341-343, 1986.

[8] Kriegesmann, "Microstructure Control during Consolidation of Fine Grained Recrystallized Silicon Carbide", Key Engineering Materials, vol. 264-268, pp. 2199-2202, 2004.
[9] Stobbe and Hack, "Porous ceramic body and method for production thereof", US7699903 BB.
[10] Kriegesmann, "Finely layered recrystallized silicon carbide for diesel particulate filters", CFI Ceramic Forum International vol. 10, pp. E25-E29, 2004.
[11] Darvishzadeh, Priezjev, "Effects of crossflow velocity and trans-membrane pressure on microfiltration of oil-in-water emulsions", Journal of Membrane Science, vol. 4230-424, pp. 468-476, 2012.
[12] DIN 19643-1 "Treatment of water of swimming pools and baths - Part 1: General requirements", 2012.

FABRICATION OF POROUS CERAMICS WITH CYLINDRICAL PORES AND INCORPORATING PORES BY UNIDIRECTIONAL SOLIDIFICATION PROCESS

Shunkichi Ueno and Jun-Woo Lee
College of Engineering, Nihon University, Tamura-machi, Koriyama, Fukushima 963-8806, Japan

ABSTRACT

In this paper, a fabrication process of porous ceramics with cylindrical pores and incorporating pores was discussed. The cylindrical pores were formed during the unidirectional solidification of aqueous slurry containing CO_2 gas under reduced pressure and a composite of ice and ceramics powder was obtained. An incorporating porous structure consisted of ceramic powder was obtained by vacuum freeze drying process. The porous green body with cylindrical pores and incorporating pores was sintered at 1250°C for 5 hours. The porosity and pore size was well controlled by the pressure during the solidification and sintering temperature and time.

INTRODUCTION

In the recent ceramics engineering, such as refractory, the amount of the binder as a sintering agent is increasing to obtain a high density bulk. To release the exhaust gas during the debinding process, the porous structure of the refractories must be controlled high porosity and high strength. In terms of porous structure, a cylindrical pore for fluid permeability and fine incorporating pores for controlling the pressure loss are necessary for the porous refractories. Furthermore, it is desired the refractories have high specific strength. Since the pores with 100 m diameter are able to be a fracture origin of oxide ceramics, the pores size in these porous refractories must be controlled within 100 m.

The specific strength of lotus-type porous steel with number of aligned cylindrical pores shows a constant value[1]. Hence, the shape of the porous structure for lotus type porous metals and/or ceramics[2] is suitable for the porous refractories. However, since the lotus-type porous metals and/or ceramics are fabricated by unidirectional solidification under pressurized hydrogen gas, no incorporating pores are formed and the pore size for lotus type porous ceramics is larger than 100 m[2]. Furthermore, the porosity of the lotus-type porous ceramics is equal to or lower than 40%. The porous refractories required high porosity more than 80%.

The fabrication method for a high porosity porous ceramics is reported by several research groups using freeze drying method[3,4].

Therefore, a new fabrication method for a porous ceramics with high porosity, directional pores, fine incorporating pores and high specific strength is required. In this paper, we propose a fabrication method of a porous ceramics with high porosity and with cylindrical pores and incorporating pores[5]. The fabrication process of the porous ceramics included unidirectional solidification of slurry, vacuum drying and sintering processes.

EXPERIMENTAL

First, the formation mechanism of cylindrical pores in ice was examined using water containing CO_2 gas. The experimental procedures can be found in our previous repot[5] in detail. CO_2 gas flowed into distillated water for 30 minutes then the water poured into the mold. The bottom of the mold and the side wall of the mold are consisted with copper plate and clear acrylic plate as shown in Figure 1. The bottom of the mold is placed in contact with cold ethanol. The temperature of the ethanol was controlled to -50°C. The unidirectional solidification of the water occurs from the bottom of the mold. The pressure during the solidification was controlled by the height of the water as shown in Fig.1.

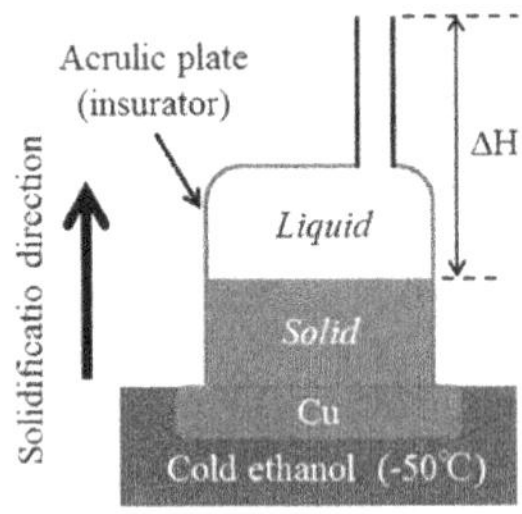

Figure 1. The configuration of the mold using the unidirectional solidification of water[5], J-W. Lee et al., Journal of the Technical Association of Refractories Jpn. 34, 226-230 (2014). ©2014, the Technical Association of Refractories Jpn.

For preparation of porous ceramics, unidirectional solidification of aqueous slurry with CO_2 gas was performed. The powder of Okubo pottery stone which come out of Fukushima prefecture was used in this experiment. The main component of this powder is feldspar. 10 g of feldspar powder and 0.2 g of PVA (Polyvinyl alcohol) binder was added in distillated water. CO_2 gas flowed into the aqueous slurry at room temperature for 30 minutes then the slurry poured into the mold. The bottom of the mold and side wall of the mold is consisted of copper plate and heat insulator. The mold was placed in vacuum chamber as shown in Figure 2. The bottom of the mold is placed in contact with the chiller cooled by ethanol. The temperature of the ethanol was controlled to -50°C. The unidirectional solidification of the water occurs from the bottom of the mold. The pressure during the solidification was controlled in reduced pressure by the aspirator.

The solidified bulk with ice and oxide powder was dried by freeze drying method for 10 hours. Since a high density bulk can be obtained using the feldspar powder by the sintering at 1250°C in industry, the sintering temperature in this experiment fixed to 1250°C. The porosity of the sintered sample was measured by Archimedian method.

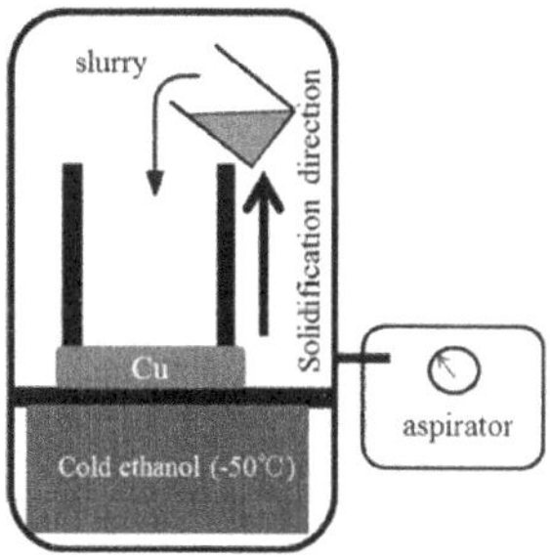

Figure 2. The configuration of the mold using the unidirectional solidification of slurry under reduce pressure[5], J-W. Lee et al., Journal of the Technical Association of Refractories Jpn. 34, 226-230 (2014). ©2014, the Technical Association of Refractories Jpn.

RESULTS AND DISCUSSION

Murakami et al. reported that cylindrical pores are formed by unidirectional solidification of soda water[6]. Figure 3 (a) and (b) show the ice with cylindrical pores solidified under 0.10 and 0.11 MPa. The pore diameters for these samples are 70.97 and 37.10 m, respectively. Since no porous structure was formed near the bottom of the mold, the pore diameter was measured at 4.4 mm upper from the bottom of the mold. The same tendency can be found in the fabrication of porous metals by casting method[1,7]. The pore diameter drastically decreased with slight increasing pressure. Figure 4 shows the pores diameter of the ice against the pressure. Where, d^3 dimension, namely, the dimension of volume of pores was plotted against the pressure. The volume of pores decreased with increasing pressure. The same tendency can be found in the case of lotus-type porous metals and/or ceramics those are fabricated by unidirectional solidification under pressurized hydrogen gas[1,2,7]. Hence, the formation mechanism of cylindrical pores for this sample can be explained the same to that of lotus-type porous metals and/or ceramics. In the formation mechanism of lotus-type porous metals, an excess hydrogen atom in solid phase around the solid-liquid interface diffuse into pores during the unidirectional solidification as shown in Figure 5. In this figure, the tops of pores contact with liquid phase and the liquid phase contacts with atmospheric gas. Hence, the pressure of inner pores increased with increasing the external pressure according to equation (1). According to Boyle's law, the volume of pores decreased with increasing external pressure.

$$P = \frac{T}{R} + P_0 \quad (1)$$

Where, P, T, R and P_0 denote the pressure of inner pores, surface tension of the liquid phase, radius of curvature for pore-liquid interface and external pressure, respectively.

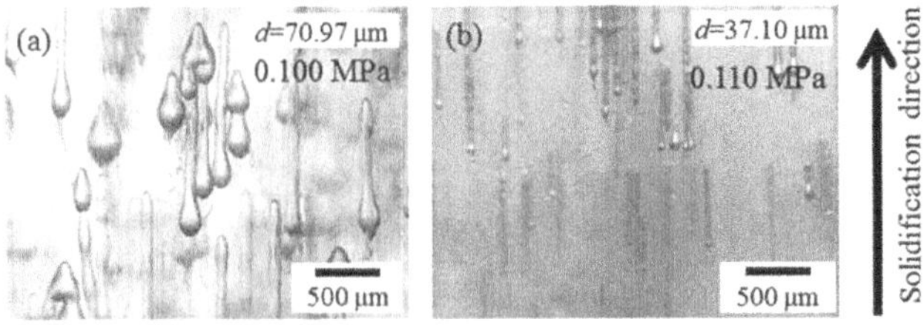

Figures 3. The ice with cylindrical pores solidified under 0.10 and 0.11 MPa[5], J-W. Lee et al., Journal of the Technical Association of Refractories Jpn. 34, 226-230 (2014). ©2014, the Technical Association of Refractories Jpn.

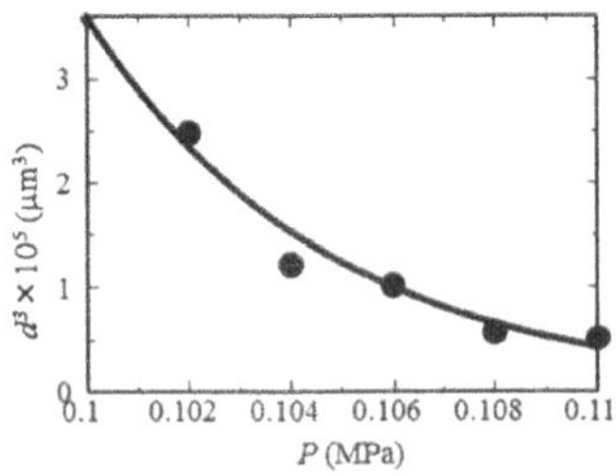

Figure 4. The pores diameter in the ice solidified[5], J-W. Lee et al., Journal of the Technical Association of Refractories Jpn. 34, 226-230 (2014). ©2014, the Technical Association of Refractories Jpn.

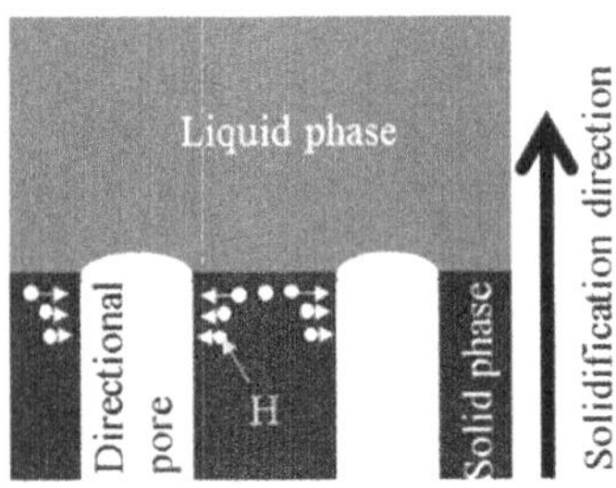

Figure 5. The formation mechanism of cylindrical pores for lotus-type porous metals and/or ceramics in pressurized hydrogen atmosphere[5], J-W. Lee et al., Journal of the Technical Association of Refractories Jpn. 34, 226-230 (2014). ©2014, the Technical Association of Refractories Jpn.

In the formation of cylindrical pores for lotus-type porous metals, the pore diameter can be controlled by the pressure from 0.1 to 1.0 MPa[7]. On the other hand, in this experiment, the pore size drastically decreased by a small pressure range from 0.10 to 0.11 MPa. Since the pressure of inner pores can be expressed by equation (1), the surface tension for the liquid phase is closely related to the pore formation. The surface tension of molten metals is larger than that of water. For example, the surface tension of molten copper at the melting point is 1306 dyne/cm[0]. On the other hand, the surface tension of water at 20°C is 72 dyne/cm. According to equation (1), the effect of external pressure to the pressure of inner pores in the case of molten metals is smaller than that of water. Hence, in the case of pore formation in ice, it can be considered that a small pressure range is necessary to control the pore diameter.

These results suggest that an accurate control of the external pressure during the solidification is required to fabricate the porous ceramics with cylindrical pores.

On the basis of above results, the porous ceramics with cylindrical pores and fine incorporating pores was prepared by unidirectional solidification under reduced pressures, freeze drying and sintering process. The sintering temperature and time was fixed to 1250°C and 5 hours. The pressure during the solidification was controlled in 0.10, 0.07, 0.05 and 0.03 MPa. The experimental conditions and porosity for each sample was summarized in Table 1. Figure 6 show the transverse cross section view of the sintered samples. For the samples of C and D, many small circular pores can be observed. Figures 7 show the longitudinal cross-section view of the sintered samples. For the samples of C and D, many directional pores can be seen. Hence, these directional pores possess cylindrical shape. The shrinkage ratio for all samples was around 40%. In the case of samples A and B, no cylindrical pores are formed. Since the weight of slurry is larger than that of water, the pressure at liquid-pore interface during the solidification is larger than 0.10 MPa. Hence, no cylindrical pores are formed under 0.10 MPa. Figure 8 shows the porosity of the samples. The porosity of the samples increased with decreasing pressure below 0.07 MPa. Since the all of samples has incorporating pores that were formed by evaporated of ice from ice/powder composite during vacuum freeze drying process, the porosity of 60% is contributed to the porosity for all samples. On the other hand, the increase of the porosity for the samples prepared below 0.07 MPa is correspond to the formation of cylindrical pores. The formation of cylindrical pores contributed to the increase of porosity of the sample.

In the present work, slurry concentration and sintering conditions were fixed. However, the porosity and pore size of the samples can be controlled by slurry concentration, pressure during the solidification, sintering temperature and sintering time. Furthermore, the porosity of cylindrical pores can be controlled by the pressure during the solidification of slurry.

Table 1. The experimental condition for each sample.

Sample	Pressure during the solidification	Porosity
A	0.10 MPa	60.8%
B	0.07 MPa	60.9%
C	0.05 MPa	70.6%
D	0.03 MPa	77.8%

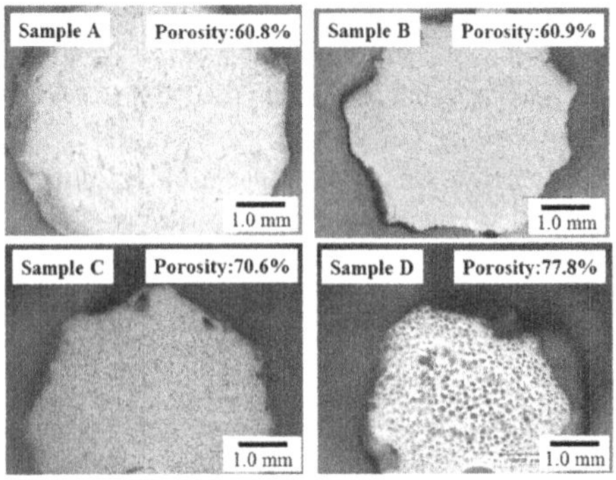

Figures 6. The transverse cross section view of the sintered samples[5], J-W. Lee et al., Journal of the Technical Association of Refractories Jpn. 34, 226-230 (2014). ©2014, the Technical Association of Refractories Jpn.

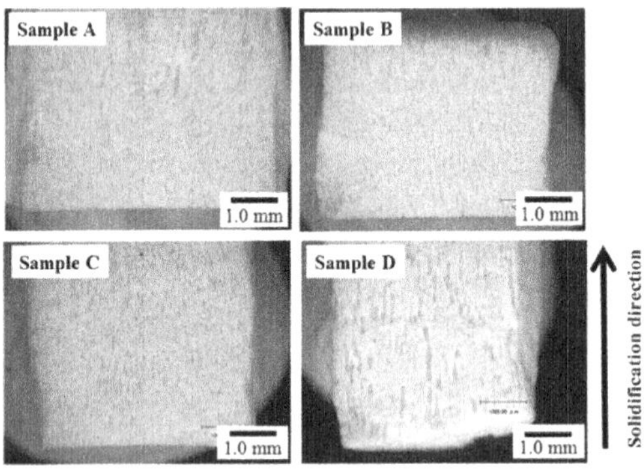

Figures 7. The longitudinal cross-section view of the sintered samples[5], J-W. Lee et al., Journal of the Technical Association of Refractories Jpn. 34, 226-230 (2014). ©2014, the Technical Association of Refractories Jpn.

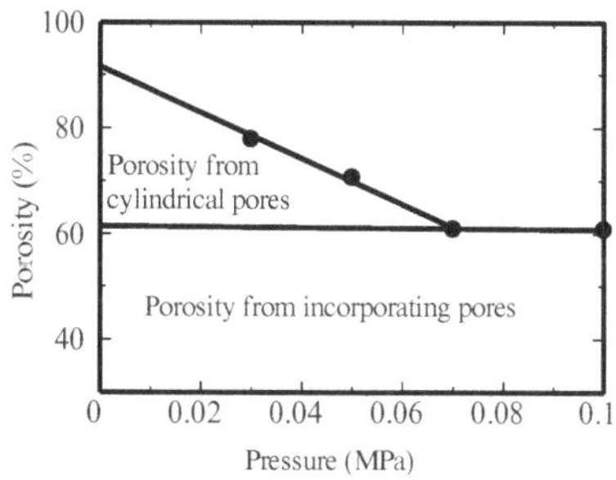

Figure 8. The porosity of the samples[5], J-W. Lee et al., Journal of the Technical Association of Refractories Jpn. 34, 226-230 (2014). ©2014, the Technical Association of Refractories Jpn.

The fabrication method for a porous ceramic with cylindrical pores and fine incorporating pores proposed in this paper can be applied to different porous ceramics. For example, a porous silicon carbide ceramics with bi-modal pores in shape is desirable structure for heat sink for power devices. The author tried to prepare the porous silicon carbide with bi-modal pores in shape using the same fabrication method. Figure 9 shows an example of the porous silicon carbide ceramics. Where, the sintering of the green body was performed by microwave sintering. The authors will provide the experimental procedures and the formation mechanism in detail in future.

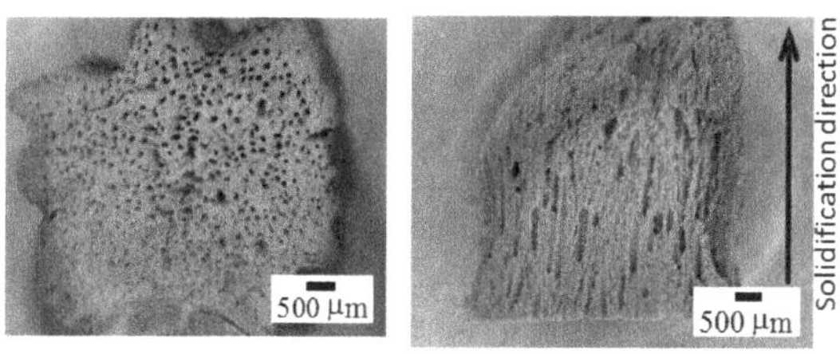

Figure 9. An example of porous silicon carbide with bi-modal pores in shape.

CONCLUSION

A fabrication method for a porous ceramic with cylindrical pores and fine incorporating pores was proposed. The formation mechanism of cylindrical pores was examined. From the experimental results, it can be concluded as follows;

(1) The formation of cylindrical pores during the solidification of water containing CO_2 gas can be controlled by a small pressure range from 0.10 to 0.11 MPa.

(2) The formation of cylindrical pores during the solidification of the slurry containing CO_2 gas occurs under reduced pressure.

ACKNOWLEDGMENT

This work was supported by a grant of The Technical Association of Refractories, Japan in 2013. And this work was performed under the Cooperative Research Program of "Network Joint Research Center for Materials and Devices" in 2014.

REFERENCES

[1] S.K. Hyun, K. Murakami and H. Nakajima, "Anisotropic mechanical properties of porous copper fabricated by unidirectional solidification" *Materials Science and Engineering*, **A299**, 241-248, 2001.

[2] S. Ueno, L.M. Lin and H. Nakajima, "Formation Mechanism of Porous Alumina with Oriented Cylindrical Pores Fabricated by Unidirectional Solidification", *Journal of the American Ceramic Society*, **91**, 223-226, 2008.

[3] M. Fukushima, M. Nakata, Y. Zhou, T. Ohji and Y. Yoshizawa, "Fabrication and properties of ultra highly porous silicon carbide by the gelation–freezing method", J.Eur.Ceram.Soc., 30, 2889-2896, 2010.

[4] T. Ohji and M. Fukushima, "Macro Porous Ceramics; Processing and Properties", Int. Mater. Rev., 57[2], 115-131(17), 2012.

[5] J-W Lee, T. Akatsu and S. Ueno, "Fabrication of Porous Ceramics with Bi-modal Pores in Shape by Unidirectional Solidification of Aqueous Slurry Containing CO_2 Gas", *Journal of the Technical Associations of Refractories, Japan*, **34**, 226-230, 2014.

[6] K. Murakami and H. Nakajima, "Formation of Pores during Unidirectional Solidification of Water Containing Carbon Dioxide", *Materials Transactions*, **43**, 2582-2588, 2002.

[7] H. Nakajima, "Fabrication, properties and application of porous metals with directional pores", *Progress in Materials Science*, **52**, 1091-117, 2007.

[8] A. Kasama, T. Iida and Z. Morita, "Temperature Dependence of Surface Tension of Liquid Pure Metals", *Journal of Japan Institute of Metal Materials*, **40**, 1030-1038, 1976.

Advanced Sensors for Energy, Environment, and Health Applications

PRINTED CANTILEVERS AND MOS GAS SENSORS FOR HAZARDOUS GAS DETECTION AT ROOM TEMPERATURE

Hélène Debéda [1], Van Son Nguyen [1], Fernando Almazán[2], Maria Pina Pilar [2], Véronique Jubéra [3], Claude Lucat [1]
[1] Université de Bordeaux, Laboratoire IMS, UMR 5218, Talence, France
[2] Nano Institute of Aragon, Zaragossa, Spain
[3] CNRS, ICMCB, UPR 9048, F-33600 Pessac, France

ABSTRACT

Among numerous technologies nowadays available for thick-films fabrication in microelectronics, the very simple, flexible and cheap screen-printing equipment allows deposition of a large choice of materials on various substrates. Modifications of standard screen-printing is moreover proposed for a few decades for integration purpose, performance improvement, lower temperature processing and even for MEMS applications by addition of a sacrificial layer process. In this work, the feasibility of a new generation of screen-printed gas sensors is undertaken. They are based on either gas sensitive nanoparticles or on free-standing cantilevers coated with a sensitive layer to achieve room temperature gas sensors. Hydrogen detection is performed at room temperature using the current variation of ZnO nanoparticles-based layers, printed on interdigitated Pt electrodes and then sintered at 400°C. Moreover, a strontium carbonate sacrificial layer process is implemented to release free-standing elements for MEMS fabrication. Resonators consisting of piezoelectric Au/PZT/Au cantilevers are successfully realized. Coated with inorganic zeolite, these cantilevers are used for benzene detection at room temperature. In this case, a high temperature sintering process (900°C) is necessary to provide good piezoelectric properties.

INTRODUCTION

Low-cost sensors with high performances and low power consumption are still a challenge. Metal Oxide Semiconductor (MOS) belong to the most developed gas sensors because of a simple change of their electrical resistivity with the surrounding atmosphere. They are generally operated at high temperature (200-400°C) and thus may induce excessive power consumption, undesirable long term drift problems and possible fire hazard. To reduce the operating temperature while keeping a good sensitivity, use of nanomaterials is subject of many researches since enhancement of the gas-surface interaction is expected. Indeed, the sensitivity increases when the grain size is decreasing[1-3].

Likewise, microcantilever-based chemical sensors are widely investigated because of their high sensitivity at room temperature[4-7]. To perform detection of chemical species, in gas or liquid media, the microcantilever is usually coated with a chemically sensitive layer that aims to selectively sorb the target analyte. The sorbed species modify the mechanical properties of the sensitive coati. The mass, the rigidity and the surface stress of the microstructure are therefor affected. The sorbed species can consequently be detected either by measuring the microcantilever's resonant frequency shift (dynamic mode) or the microcantilever's quasi-static deflection (static mode). Polymers are often used as sensitive layers for gas detection (e.g. VOCs). More recently, silicon cantilevers have been functionalized with inorganic coatings like zeolite[8-9]. In this paper, we will focus on the performance of resistive sensors ZnO MOS sensor based on nanomaterials for hydrogen detection at room temperature. Hydrogen is widely used as an energy carrier and as a chemical reactant and whose detection is essential to alert the formation of potentially explosive mixtures with air and to help for the prevention of the explosion risk[10]. On the other hand, screen-printed piezoelectric cantilevers platforms coated with de-aluminated large pore zeolites will be used as chemical sensors for environmental benzene detection at room temperature. Benzene is becoming of major importance due to its significant adverse effects on human health[11].

The screen-printing processing of the resistive MOS sensors and of the cantilever-based sensors will be first detailed before gas tests. Performances of the sensors will be further presented.

GAS SENSOR PROCESSING

Among numerous technologies nowadays available in microelectronics, screen-printing belongs to the most simple, flexible, reliable and cheapest way for thick-film deposition. This technology, mainly used in microelectronics for interconnections and packaging purposes[12], has been extended to the fabrication of passive components such as sensors[13] and more recently to inorganic material-based MEMS or polymer-based microfluidic or micromechanical devices by using a sacrificial layer process[14-16]. The choice of such technology for sensors and MEMS fabrication is justified because of the large choice of substrates, starting materials, layers geometries, multi-layers deposition and the control of the layer's porosity. MOS sensors based on ZnO nanomaterials are fabricated using the standard screen-printing technology. On the other side, piezoelectric cantilevers based sensor are realized thanks to the association of the standard screen-printing technology with a sacrificial layer process.

The design of ZnO nanoceramics gas sensor is shown in figure 1. The sensor consists of $6x8x0.25mm^3$ alumina substrates with platinum interdigitated electrodes (Pt IDTs) on one side and platinum heater (meander shape) on the other side, connected on a socket with Pt wire. The interdigitated electrodes and heater are printed and fired at 950°C using a Pt5545 commercial paste from Electroscience Laboratory. The sensitive layer is screen-printed on the Pt IDTs with commercial ZnO nanoparticles Nanotek® of 40-100nm grain having a specific area ($10\text{-}25m^2/g$) well adapted for gas sensing. The semiconductor paste is prepared with the ZnO nanoparticles blended with 67%wt of an organic binder (ESL 400). After a manual mixing in a mortar, the nanoparticles dispersion is optimized thanks to a three-roll mill. Once printed, the layers are dried 20 min at 120 °C and fired 30 min at 400 °C for a complete removal of organic binder with respects to the TGA analysis (figure 2a). AFM characterization of the ZnO layers before and after thermal treatment show that the ZnO nanoparticles have a quite homogeneous size distribution (≈100 nm) and that the grain size is not affected by the heat-treatment (figure 2b).

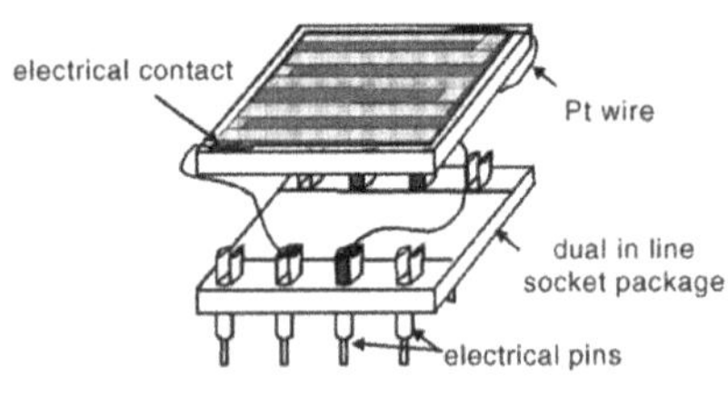

Figure 1. Scheme of the screen-printed ZnO gas sensor

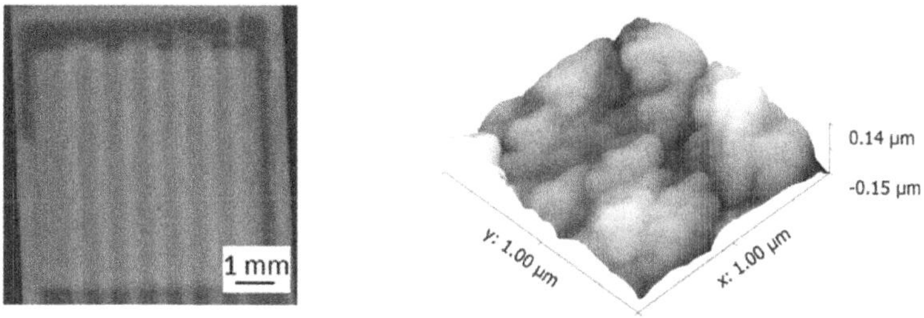

(a) (b)

Figure 2. (a) Photograph of the ZnO layer printed on the platinum IDTs and (b) AFM analysis of the printed ZnO nanoceramics.

The piezoelectric micro-cantilever based sensor (8x2x0.1 mm^3) is realized by screen-printing process associated to a sacrificial layer[14] (Figure 3a). The cantilever is composed of PZT with printed Au electrodes on each side. This symmetrical structure allows the in-plane 31-longitudinal vibration mode. A Pt microheater is deposited on the Al_2O_3 substrate and can be used for species desorption. For the cantilever fabrication, the Pt microheater is first screen printed on the alumina substrate, dried 20min at 120°C and fired 10 min at 850°C. Afterwards, to obtain the cantilever, a sacrificial layer based on epoxy and $SrCO_3$ is deposited on the substrate and polymerized 20min at 120°C. Then bottom Au electrode, PZT layer and top Au electrode are printed successively with a drying step at 120°C during 20min between each deposition. To improve the densification before firing, the dried layers are isostatically pressed 1min at 1kbar. The samples are then fired 2 hours at 900°C under air atmosphere. Finally, dissolution of the sacrificial layer is performed in the 0.9$mole^{-1}$ H_3PO_4 aqueous solution. Au/PZT/Au cantilevers are then poled at 280°C with an electric field of 5$kVcm^{-1}$.

For gas sensing, the PZT cantilevers are functionalized with zeolite. One of the main features of zeolite type materials is the tunable sorption properties by adjusting their Si/Al ratio and their exchange cations. The Si/Al determines to a large extent the hydrophobic/hydrophilic and polar/nonpolar character of the interactions between the zeolite and the fluid phase molecules. Moreover, advantages of large pore zeolites with low Al content over activated carbon have been previously demonstrated for VOC adsorption in presence of water[17]. Commercial powder of faujasite type zeolites with 1-3 µm grain size (Zeolyst®) are dispersed in ethanol (1%wt) and drop-coated on the PZT cantilevers. This framework topology, with 3-dimensional channel system with pore opening 7.35 Å in size, has been chosen on the basis of its capabilities for BTX adsorption[18]. In particular, different Y type zeolites with Si/Al ratio varying from 15 to 40, and extra-framework cation (H+, Cu++) are studied. Figure 3b shows a photograph of the zeolite DAY (Si/Al=40, proton H+ as compensation cation) coatings, once ethanol solvents are evaporated by using the Pt microheater. Three amount of zeolites (228, 571 and 808 µg) have been deposited on the cantilever.

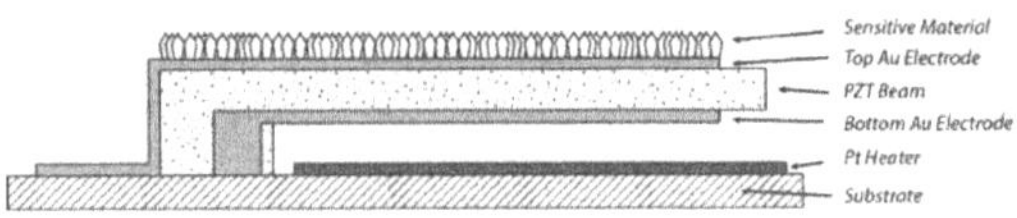

(a)

(b)

Figure 3. (a) Scheme of the Au/PZT/Au cantilever (b) top view of the Au electrode coated with zeolite

HAZARDOUS GAS DETECTION

Resistive ZnO sensors: hydrogen detection

Hydrogen detection with ZnO layers is performed in a test chamber (volume ≈ 1.7 cm^3) equipped with a quartz window, thus enabling sample illumination. The illumination is made with an halogen light (UV intensity 15 $\mu W/cm^2$) to reach a measurable current (figure 4a) during the latter gas tests. A Labview environment controls the gas mixture of the carrier gas (dry air) and hydrogen delivered from a calibrated gas cylinder (5% H_2 / 95% N_2). The total flow is 1000 ml/min for a selected hydrogen concentration of 300 ppm. Before starting the measurements, the sensors are stabilized 2 hours at 200 °C under light illumination using the Pt microheater. Responses S (%) = (I_{air} - I_{H2}) / I_{air} of the ZnO printed sensors to 300 ppm of H_2 under dark and halogen light at room temperature are shown on figure 4b. The responses at room temperature and at 150°C are reported in the table I. One can see the very low value of S when T=22 °C though the response obtained with halogen light at room temperature is equivalent to those observed under dark when T = 150 °C. The higher sensitivity of the ZnO layer is observed under halogen light at 150°C.

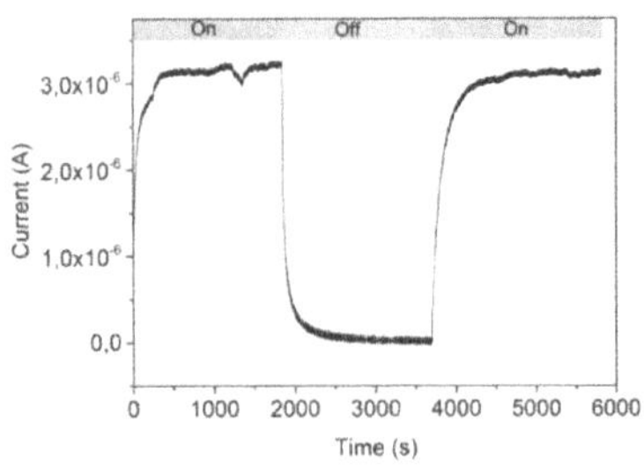

(a)

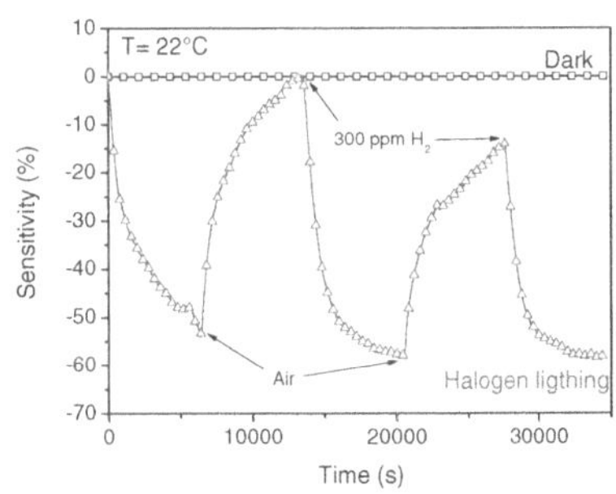

(b)

Figure 4. Current evolution of the ZnO printed film at 22°C (bias voltage 1V): (a) in dry air with ('On') and without halogen lighting ('Off'), (b) under dry air or 300ppm H_2

Table I: Gas response to H_2 of ZnO printed sensor at different temperatures

Light source	Temperature (°C)	$\lvert S(\%)\rvert = \lvert I_{air} - I_{300ppmH2}) / I_{air}\rvert$
Dark	22	0
	100	0
	150	38-40
	200	37
Halogen lighting	22	44-57
	100	16
	150	71-78
	200	37

Cantilever-type sensors: benzene detection

The functionalized PZT cantilevers are placed in a gas chamber (35 cm^3) under a controlled gas flow (100ml/min) of dry air. The desired benzene concentration levels fed to the chamber (ranging from 1 to 1000ppm) are attained by means of calibrated cylinders and a set of mass-flow controllers. During the experiments, the impedance of the functionalized cantilevers is monitored using an Agilent E5061B impedance meter. The whole system is controlled under a LabVIEW environment. The resonance frequency evolution as a function of time is deduced from this measurement. The measured sensitivities for the first 31 longitudinal mode are 0.064 , 0.191 and 0.306 $Hz.ppm^{-1}$ for PZT cantilevers coated respectively with 228, 571 and 808 μg of DAY (Si/Al=40, proton as compensation cation)(Figure 5a). The sorption affinity towards benzene for two type of DAY acid zeolites with Si/Al ratio of 15 and 40 and their copper exchanged forms is also evaluated on Si microcantilevers[20] coated with discrete crystals (size of few μm) of the aforementioned materials (Figure5.b). As it was expected, the higher Si/Al ratio, higher is the sorption affinity (from 2.4 % wt to 4.3% wt). Moreover, for a given Si/Al ratio (40), the Cu-exchanged form allows to increase benzene uptake in 152 percentage points over the protic form.

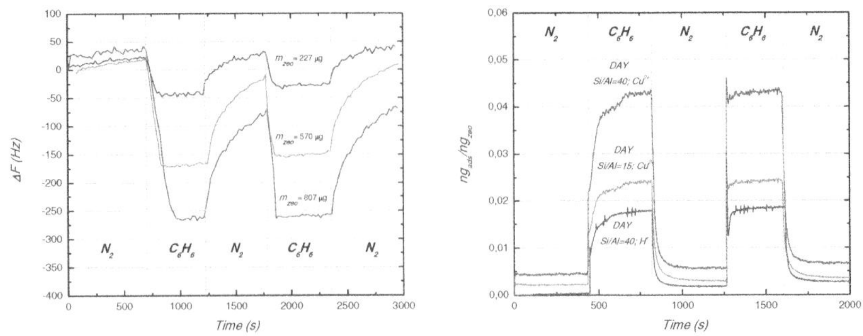

(a) (b)

Figure 5. Sensor response under benzene (1000 ppm) with: (a) screen-printed PZT cantilevers with increasing DAY zeolites masses (Si/Al=40; H^+); (b) Si microcantilevers coated with different DAY (change of Si/Al ratio and extra-framework cation).

CONCLUSION

Measurements on MOS ZnO sensors show clearly that the halogen light providing a UV light intensity of 15 $\mu W/ mm^2$ gives enough photo-activation for gas sensitivity improvements under hydrogen at room temperature. The sensitivities of ZnO printed nanoceramics compare well to those of the literature obtained for few hundred ppm of H_2 and with different ZnO morphologies (1D to 2D) and size particles[21-26]. In addition, a low-cost fabrication process has been proposed for screen-printed gas sensors based on ZnO nanoparticles sintered at only 400°C. Moreover, this ZnO nanoceramics could be integrated on flexible substrates like polyimide[27].

Concerning PZT cantilevers coated with zeolites, the preliminary results have paved the way for new approaches under study to fabricate low cost-portable-outperforming sensors for BTX monitoring. The performances (sensitivities and selectivity) could be enhanced by improving the sensor response (ratio signal/noise) and by choosing the appropriate zeolite type. Furthermore, an incorporation on top of the PZT electrodes of macro-meso-microporous sensing layers based on zeolites by using appropriate screen printing pastes is under progress.

ACKNOWLEDGMENTS

Région Aquitaine (CTP) and the French national agency for research (ANR) have supported these researches.

REFERENCES

[1]N. Yamazoe et al, "Receptor function and response of semiconductor gas sensor", Journal of sensors, 21 pages, Review Article, 2009

[2]B.P.J. de Lacy Costello et al, "Highly sensitive room temperature sensors based on the UV-LED activation of zinc oxide nanoparticles", Sensors and Actuators B, 134 945-52, 2008

[3]S. Mishra et al, "Detection mechanism of metal oxide gas sensor under UV radiation", Sensors and Actuators B, 97 387-90, 2004

[4] K.M. Goeders et al, "Microcantilevers: Sensing Chemical Interactions via Mechanical Motion", Chemical Reviews, 108, 522-542, 2008
[5]A. Boisen et al, "Cantilever-like micromechanical sensors", Reports on Progress in Physics, 74, 1-31, 2011
[6]X. Li et al, "Integrated microcantilevers for high-resolution sensing and probing", Measurement Science and Technology, 23, 022001, 40pp, (2012).
[7]M. Sepaniak et al, "Microcantilever Transducers: A new Approach in Sensor Technology", Analytical Chemistry, 74, 568-575, 2002
[8]M. A. Urbitzondo et al, "Zeolite-modified cantilevers for the sensing of nitrotoluene vapors", Sensors and Actuators B, 137, pp. 608–616 (2009).
[9]M. A. Urbitzondo et al, "Detection of organic vapours with Si cantilevers coated with inorganic (zeolites) or organic (polymer) layers", Sensors and Actuators B, in press, 201).
[10]T. Hübert et al, "Hydrogen sensors – A review", Sensors and Actuators B: Chemical, 157, 329-52, 2011
[11] D.A. Savitz et al,"Review of epidemiologic evidence on benzene and lymphatic and hematopoietic cancers", American Journal of Industrial Medicine, 31(3), pp.287-295, 1997
[12]M. Haskard et al, "Thick film technology and applications", Electrochemical Publications, 1997
[13]F.Ménil et al, "Screen-printed thick-films : from materials to functionnal devices", Journal of European Ceramic Society, 25, pp.2105-2113, 2005
[14]C. Lucat et al, "Microsystems elements based on free-standing thick-films made with a new sacrificial layer process", Microelectronics Reliability, 48, 6, pp. 872-875, 2008
[15]N. Serra et al, "Fabrication of polymer-based micro devices: formulation and study of the paste," Sensors and actuators A, 172, pp.336-340, 2011
[16]S.Kok et al, "Fabrication and characterization of free-standing thick film piezoelectric cantilevers for energy harvesting", Measurement Science and Technology, 20, 124010 (13pp), 2009
[17]El Brihi, "Determining volatile organic compounds' adsorption isotherms on dealuminated Y zeolite and correlation with different models », J. Chem. Eng. Data 47, pp 1553-1557, 2002
[18]M. Yu et al." Adsorption of Benzene Mixtures on Silicalite-1 and NaX Zeolites", Microporous and Mesoporous Materials, 96, 376 – 385, 2006.
[19] P.Clément et al, " Screen-printed PZT cantilevers coated with nano and micro powders for benzene detection at room temperature", 3rd International Conference on Materials and Applications for Sensors and Transducers, IC-MAST, Prague, 13-17 September 2013
[20] M.P. Pina et al, "Explosives detection using nanoporous coatings", SPIE Proceedings. Proc. SPIE 8031, 803124 doi:10.1117/12.883780, 2011
[21] M. Tonezzer et al, "H_2 sensing properties of two-dimensional zinc oxide nanostructures", Talanta, 122 201-8, 2014
[22]O. Lupan et al, "Selective hydrogen gas nanosensor using individual ZnO nanowire with fast response at room temperature", Sensors and Actuators B 144 56–66, 2010
[23] L. C. Tien et al , "Hydrogen sensing at room temperature with Pt-coated ZnO thin films and nanorods", Appl. Phys. Lett. 87 222106, 2005
24A.Qurashi et al, "Ultra-fast Microwave Synthesis of ZnO Nanowires and their Dynamic Response toward hydrogen gas", Nanoscale Res. Lett.4 948–954, 2009
[25] Q. Ahsanulhaq et al, "Electrical and gas sensing properties of ZnO nanorod arrays directly grown on a four-probe electrode system", Electrochemistry Communications, 12 475–478, 2010
[26] Z.H. Lim et al, "A facile approach towards ZnO nanorods conductive textile for room temperature multifunctional sensors", Sensors and Actuators B Chemical 151 121–126, 2010
[27] D. Zappa et al, "Zinc oxide nanowires deposited on polymeric hotplates for low-power gas sensors", Procedia Engineering 47 1137 – 1140, 2012

SENSING CHARACTERIZATION OF THE MOS MICRO GAS SENSOR ARRAY ON GAS MIXTURE

Bum-Joon Kim and Jung-Sik Kim*
Department of Materials Science and Engineering, The University of Seoul, Seoul, 130-743 Korea

ABSTRACT

Metal oxide semiconductor (MOS) gas sensors are very attractive owing to their low cost simplicity of use, large number of detectable gases and various potential application fields. However, the MOS gas sensor has a serious shortcoming of low selectivity in a mixture of gases, In this study MOS micro gas sensors were fabricated for detecting carbon monoxide (CO), nitrogen oxide (NO_2), ammonia (NH_3) and formaldehyde (HCHO) gases, as well as their binary mixed gas systems. Four sensing materials, Pd-SnO_2 for CO, In_2O_3 for NO_x, Ru-WO_3 for NH_3, and SnO_2-ZnO for HCHO were synthesized using a sol-gel method and deposited in the middle of sensor platform. The micro gas sensor platform was fabricated by using a MEMS technology. The sensing electrode and micro heater were designed to be a co-planar type structure with the Pt thin film layer. The gas sensitivity and sensing behaviors for gas mixture suggested that the selective adsorption of one gas with respect to others occurred for gas mixture and resulted in good selectivity for a particular gas species. Furthermore, the careful pattern recognition of sensing data obtained with sensor array makes it possible to distinguish a gas species from gas mixture and to measure its concentration.

INTRODUCTION

Metal oxide semiconductor (MOS) gas sensors are some of the most studied groups of gas sensors owing to their low cost, simplicity of use, and large number of detectable gases and various potential application fields[1]. On the other hand, MOS gas sensors have serious shortcomings of their low selectivity, response drifts and environmental influences such as temperature, vibrations and the gas flow[2]. For practical applications, MOS gas sensors have four major issues of concern: selectivity, long-term stability, reproducibility of the devices, and sensitivity[3,4].

Regarding the selectivity issue, the electronic nose (e-nose) concept has been developed to achieve the ability of classifying complex gas mixtures, such as aromas and odors, using cross-sensitive sensors[5-7]. In general, an e-nose system utilizes gas sensing signals within the sensor array, and the characteristics of individual sensors should be as diverse as possible to ensure that the partial sensor gas responses are not correlated for the reliable discrimination of a certain gas from gas mixture. As other issues related to the long-term stability and reproducibility, such e-nose systems require good reproducibility of the sensor array and high training cost for sensor maintenance. This appears to be one of key challenges requiring the breakthrough[8]. Gas sensor arrays based on a single metal oxide thin film segmented by electrodes appear to meet these requirements well because a large number of sensor devices on Si-chip have been produced using the identical fabrication process (generally CMOS and MEMS processes), and most of sensor platforms are expected to be similar. However, such multi-sensor arrays usually have uncontrollable variations during the manufacturing process. In particular, micro-electrodes for conductivity measurement often result in the substantial and unpredictable doping of sensing area with an electrode material, which alters the resistance of sensing segments in a random manner[8]. Minimizing the influence of these uncontrollable changes would provide a good step towards reproducible multi-sensor arrays. For the final issue with the gas sensitivity, the reported sensitivity of many MOS gas sensors is surprisingly high[9,10]. The current detection limits for many reducing gases are sufficient for major practical applications, making the sensitivity issue a lesser concern. On the other hand, the higher gas-sensitivity observed by nano-crystal based metal oxide layers tends to decrease during the course of exploitation[11], whereas relatively stable sensing structures, such as meso/nano-wires or stabilized crystallite films, often have comparably lower sensitivities[12,13]. Therefore, improving the stability of the sensing properties of MOS sensors is a major issue.

Gas identification techniques have attracted considerable attention over the past twenty years. The ability to monitor the leakage of combustible and explosive gases is essential for preventing accidental explosions and problems with the pollution and the toxicity. Accordingly, there is urgent demand for sensors combined with pattern recognition systems that can detect and determine the various kinds of combustible gases selectively[14,15].

In this study, micro-hotplate-based MOS gas sensors, which have a number of interesting features and are particularly attractive for their practical approach, were utilized for the detection of gas mixtures. In previous studies, we developed four different MEMS-type gas sensors for the detection of carbon monoxide (CO), nitrogen oxides (NO_x), ammonia (NH_3), and formaldehyde (HCHO). Four sensing materials with nano-sized particles for these target gases (Pd-SnO_2 nano-powder for CO, In_2O_3 nano-particle for NO_x, Ru-WO_3 nano-composite for NH_3, and hybridized SnO_2-ZnO material for HCHO) were synthesized using a sol-gel method[16-19]. Each MEMS gas sensor showed good sensing performance for its target gas, and the optimal operating temperature was determined.

In the present study, the sensing responses of these four sensors were investigated for the gas mixture, particularly binary mixed gases along with CO, NO_x, NH_3, and HCHO. Then, selectivity and response pattern for these sensors to a particular gas along mixed gases were examined.

EXPERIMENTAL

The appropriate sensing materials for four different gases; SnO_2 for CO (designated as SN sensor), In_2O_3 for NO_x (IN sensor), WO_3 for NH_3 (WO sensor), and SnO_2-ZnO for HCHO (SZ sensor), were synthesized using the sol-gel based method.[16-19] Figure 1 shows field emission scanning electron microscopy (FESEM) images of the four sensing materials. The average particle sizes were 40 nm for SN, 70 nm for IN, 1,000 nm for WO, and 20 nm for SZ sensors, respectively.

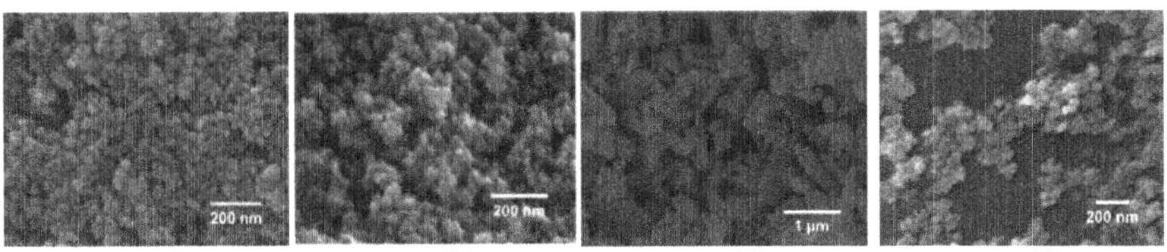

(a) (b) (c) (d)

Figure 1. Micro-structures of the four gas sensing materials: (a) Pd-SnO_2, (b) In_2O_3, (c) Ru-WO_3, and (d) Pd doped SnO_2-ZnO.

Each sol-precusor containing its sensing element was dripped with a micro-pipet on the electrode of sensor platform, and then heat-treated at its appropreate sintering temperature. Table 1 represents main features of sensing materials and the optimum temperature for their gas sensing.

Table 1. Main features of four different sensing materials

Table 1. Main features of four different sensing materials

	Composition	Average Particle Size	Optimum Temperature (°C)
SN	1% Pd-SnO_2	40 nm	225
IN	In_2O_3	70 nm	225
WO	1% Ru-WO_3	1.0 μm	367
SZ	1% Pd + SnO_2-ZnO	20 nm	367

Micro gas sensor platforms were designed with co-planar type in which sensor electrode and micro-heater were existed on the identical film (Pt thin film) layer, and fabricated using the MEMS process, previously[20]. The sensor chip size of the MEMS platform was 1.8 mm × 1.8 mm, and the membrane located in the central part of the sensor chip was 0.9 mm × 0.9 mm. Figure 2(a) shows the photograph of the IN sensor device with TO-39 package in which the sensor chip was placed and connected to the electric terminals by Au wires. The fabricated sensor had low power dissipation, and its power consumption increased linearly with increasing operation temperature as shown in Figure 2 (b). For example, power consumptions operated at 225°C for the SN sensor and 367°C for the WO sensor were 35.26 and 64.37 mW, respectively.

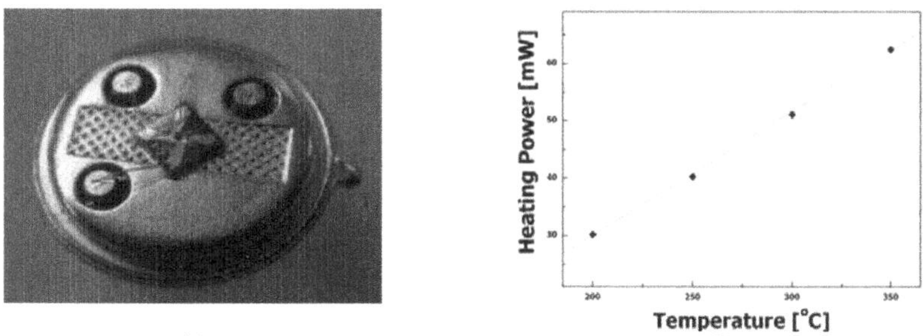

(a) (b)

Figure 2. (a) Photograph of fabricated sensor on the TO-39 package and (b) electro-thermal characteristic as the heating power vs. operating temperature.

The sensing properties were tested for gas mixture in a gas chamber in which four gas sensors were placed. The gas chamber was connected to a computer-supervised continuous gas flow system that produced the desired concentration for each gas and gas mixtures with a good reproducibility. The test gases (CO, NO_x, NH_3, and HCHO) were diluted with a nitrogen gas and carried by dry air at a constant flow rate. The total gas flow rate was about 500 ml/min. The concentration of each test gas was 0 ~ 60 ppm for CO, 0 ~ 0.6 ppm for NO_2, 0 ~ 10.0 ppm for NH_3, and 0 ~ 5.0 ppm for HCHO, respectively. Table 2 lists six gas mixtures (CO-NO_2, CO-NH_3, CO-HCHO, NO_2-NH_3, NO_2-HCHO and NH_3-HCHO) along with their mixing conditions. All sensing materials of SnO_2, In_2O_3, WO_3 and ZnO prepared in this study are typical n-type semiconducting oxides.

To quantify the sensor responses for both oxidizing and reducing gases as well as their mixtures, the gas sensitivity (S) was defined as $S = \log (R_g/R_a)$, where R_a is the sensor resistance in air and R_g is the sensor resistance after injecting the test gas. The gas sensitivity showed negative values ($S < 0$) for reducing gases, and positive values ($S > 0$) for oxidizing gases because all sensors were simultaneously sensitive to both reducing (CO, NH_3 and HCHO) and oxidizing (NO_2) gases. The gas sensing properties and selective reactions to several gases were analyzed by quantifying the sensitivity.

Table 2. Mixed gas systems for the characterization of single gases and their mixtures. (The designated numbers as # represent the test order in each binary gas mixture.)

NO_2 (ppm) \ CO (ppm)	0	30	60
0		#1	#2
0.3	#3	#4	#5
0.6	#6	#7	#8

NO_2 (ppm) \ CO (ppm)	0	30	60
0		#1	#2
5.0	#3	#4	#5
10.0	#6	#7	#8

HCHO (ppm) \ CO (ppm)	0	30	60
0		#1	#2
2.5	#3	#4	#5
5.0	#6	#7	#8

NH_3 (ppm) \ NO_2 (ppm)	0	0.3	0.6
0		#1	#2
5.0	#3	#4	#5
10.0	#6	#7	#8

HCHO (ppm) \ NO_2 (ppm)	0	0.3	0.6
0		#1	#2
2.5	#3	#4	#5
5.0	#6	#7	#8

HCHO (ppm) \ NH_3 (ppm)	0	5.0	10.0
0		#1	#2
2.5	#3	#4	#5
5.0	#6	#7	#8

RESULTS AND DISCUSSION

Two opposite cases in sensor responses were assumed before the mixed gas sensing test. One is that the test sensor shows a gas selective response to other gases, and the other is that no sensor has selectivity to all gases. Initially, gas sensor materials without selectivity might possess multi-functional adsorption sites for both gases during mixed gas flow. In an oxidizing and reducing gas mixture, both gases would adsorb on the surface of the sensing material, and the number of adsorbed gas molecules would be in the same proportion as that of each single gas. If the oxidation effect on the total sensor conductance prevails, a specific amount of oxidizing gas would counterbalance the reducing gases, and the remaining oxidizing species might contribute to the change (augment) in conductance. Therefore, a change in conductance might occur in the form of a subtraction reducing effect from oxidizing reactions. Otherwise, in the case of a mixture with reducing agents, a change in conductance might manifest as the summation of the same type reactions (reducing effects), which can have a synergic effect on the reaction of each species.

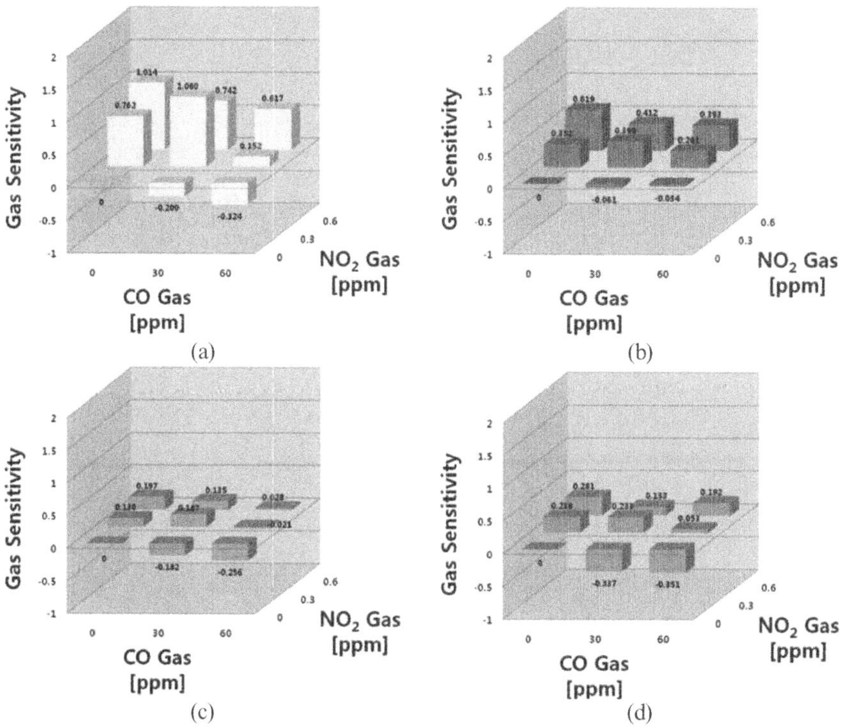

Figure 3. Gas sensing properties in the CO-NO_2 system; (a) SN, (b) IN, (c) WO, and (d) SZ sensors.

In the first system (CO-NO_2 set), which contains both reducing and oxidizing gases, the gas response to NO_2 was higher than that to CO in all sensors. Figure 3 shows the variations of the gas sensitivity of all sensors to CO and NO_2 gases and their mixtures. All sensors showed stronger responses to NO_2 gas than CO, and the sensitivity showed positive values in the presence of NO_2 gas. The SN sensor exhibited a strong response to both gases and their mixtures, whereas the IN sensor responded only to NO_2. The WO and SZ sensors exhibited similar behaviors to the SN sensor, but their sensitivities to NO_2 gas were slightly lower.

The gas sensing mechanism can be explained with depletion layer formed along the grain boundaries of MOS sensing materials. As the solid absorbs more electropositive gas (reducing gases) or electronegative gas (oxidizing gases), electrons are transferred from or to the gas, which causes a depletion layer (or space-charge layer) to be formed near the surface where the uncompensated acceptor or donor ions exist. A Schottky barrier is then formed which electrons have to overcome passing the region. Reducing gases such as CO has less electron affinity than O_2, and then the resistance decreases as the amount of CO absorbed increases. On the while, oxidizing gases such as NO_2 have stronger electron affinity than O_2 in air and the resistance increases with increasing NO_2 gas.

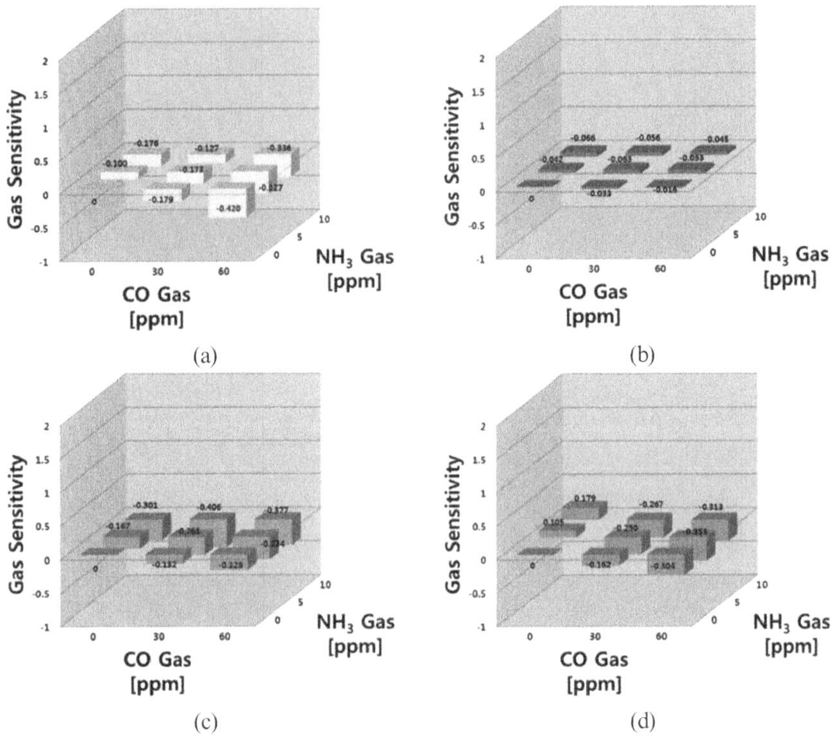

Figure 4. Gas sensing properties in the CO-NH_3 system; (a) SN, (b) IN, (c) WO, and (d) SZ sensors.

In the CO-NH_3 set, the CO gas responses were much higher than NH_3 gas in a mixture of reducing agents, but the sensitivities in the gas mixture were higher than that to each gas separately under most experimental conditions (Figure 4). The SN and SZ sensors were more sensitive to their target gas (CO) than NH_3 within the test ranges: S = -0.179 and -0.420 for SN, and S = -0.100 and -0.176 for SZ sensors (at CO 30 ppm and 60 ppm). On the other hand, the changes in resistance were slightly lower in the case of a gas mixture. The IN sensor responses were quite poor to both gases and showed a selective response to both gases. The WO and SZ sensors were sensitive to both the single gases and their mixtures, with higher sensitivities observed with the gas mixtures. For the tests in a mixture of reducing agents, the sensor responses targeting these gases did not show any synergic effects. The sensitivity of the SN sensor was -0.420 for 60 ppm CO gas, but the sensitivity was slightly lower (S = -0.327) for CO 60 ppm – NH_3 2.5 ppm. This phenomenon was not observed in the other sensors, which had gas selectivity for the specific gas species. In three mixtures (CO 60 ppm – NH_3 0 ppm, CO 60 ppm – NH_3 5.0 ppm, and CO 60 ppm – NH_3 10.0 ppm), the CO concentration was identical. As the NH_3 concentration increased, however, the sensitivity of the SN sensor was slightly lower for the gas mixtures than for the single CO gas. This suggests that the specific adsorption and selective activation of adsorption sites might occur in gas mixtures and offer a priority for the adsorption of a specific gas, which will be discussed in the following section.

The gas responses of SN, WO and SZ sensors in the CO-HCHO system showed similar behavior in the CO-NH_3 system, whereas the IN sensor showed no responses for both gases (in Figure 5). Three sensors (SN, WO and SZ) were sensitive to both the single gases and their mixtures, with higher sensitivities observed with the gas mixtures. Moreover, the SZ sensor responses toward HCHO were stronger than CO gas. Similar to the CO-NH_3 system, there was no synergic effect in the SN sensor responses to a mixture of CO and HCHO. The sensitivity of the SN sensor was -0.470 for 60 ppm CO gas, but the sensitivity was slightly lower (S = -0.429) for CO 60 ppm – HCHO 2.5 ppm. This also suggests that the specific adsorption and selective activation offer priority for the adsorption of a specific gas in the gas mixtures.

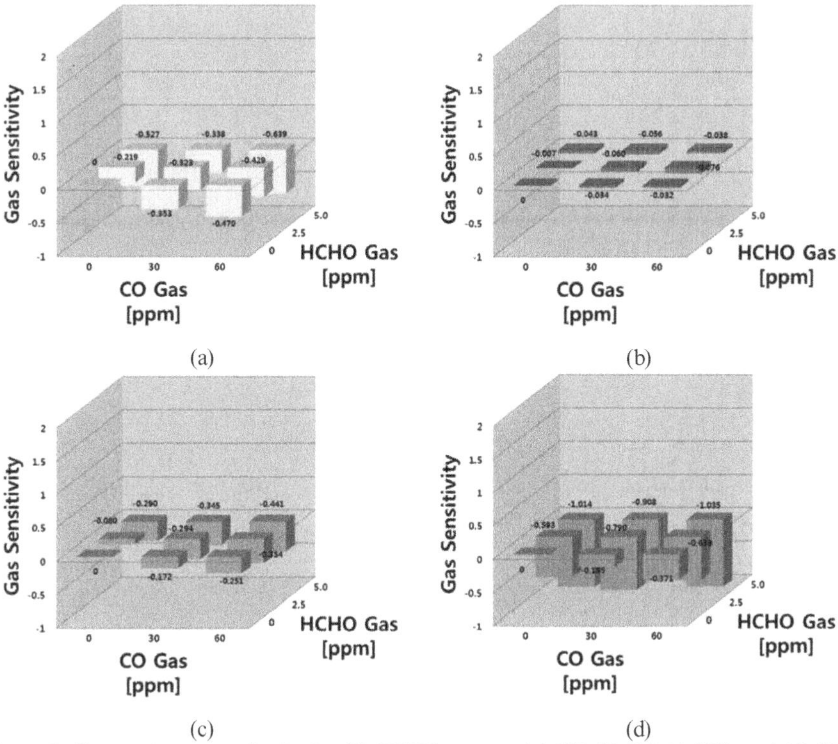

Figure 5. Gas sensing properties in the CO-HCHO system; (a) SN, (b) IN, (c) WO, and (d) SZ sensors.

In the NO_2-NH_3 system, as shown in Figure 6, the responses to NO_2 gas were stronger than those of NH_3. In the SN sensor, the sensitivities exhibited increasing behavior to NO_2 and decreasing behavior to NH_3 at higher concentrations, showing that the sensor responds to both gases (S = 0.934 at 0.3 ppm). In the gas mixtures, the changes in resistance were higher than the baseline (Ra) and decreased with increasing NH_3 concentration. On the other hand, the sensitivities to both gases had positive values (S > 0) within the test ranges. The IN sensor had a selective response to NO_2 gas but was barely sensitive to NH_3 gas. As the NH_3 gas concentration increased, the IN sensor exhibited a slight decrease in sensitivity to NO_2 gas in the NO_2-NH_3 mixed gas system. The WO sensor showed similar behavior to the IN sensor, as well as a very selective response to both gases, but it was more sensitive to NO_2 gas.

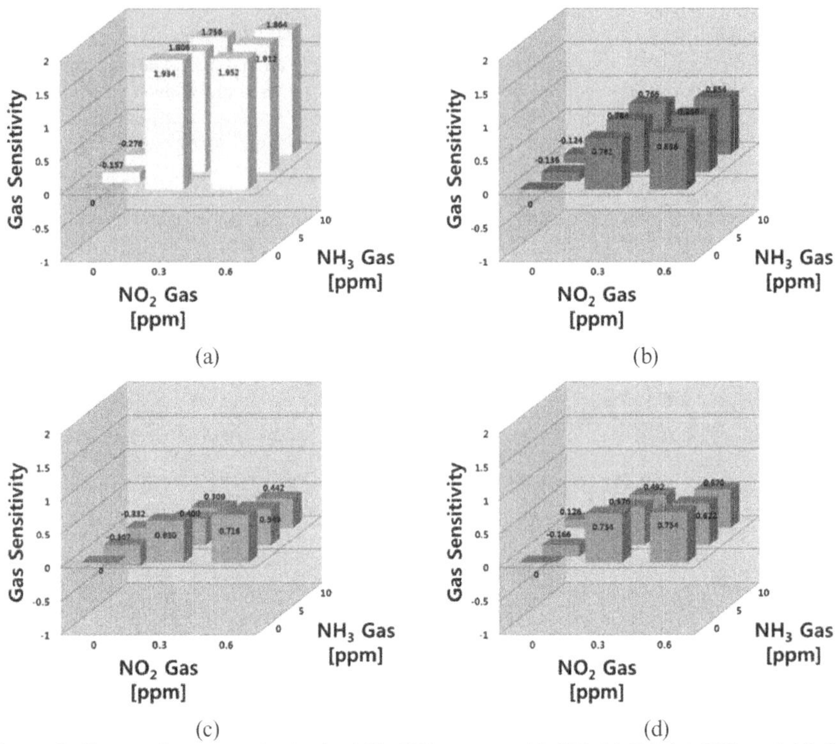

Figure 6. Gas sensing properties in the NO_2-NH_3 system; (a) SN, (b) IN, (c) WO, and (d) SZ sensors.

In the NO_2-HCHO system, the responses to NO_2 gas were stronger than those of HCHO (in Figure 7). In the SN and WO sensors, the sensitivities showed increasing behavior to NO_2 and decreasing behavior to HCHO at higher concentrations. On the other hand, the IN sensor showed a selective response to NO_2 gas, but little sensitivity to HCHO gas. As the HCHO gas concentration increased, the IN sensor showed a slight decrease in sensitivity to NO_2 gas in the NO_2-HCHO mixed gas system. The WO sensor showed similar behavior to the IN sensor, as well as a very selective response to both gases, but it was more sensitive to NO_2 gas. The SZ sensor responses toward HCHO were stronger than NO_2 gas for the NO_2 and HCHO gas mixture.

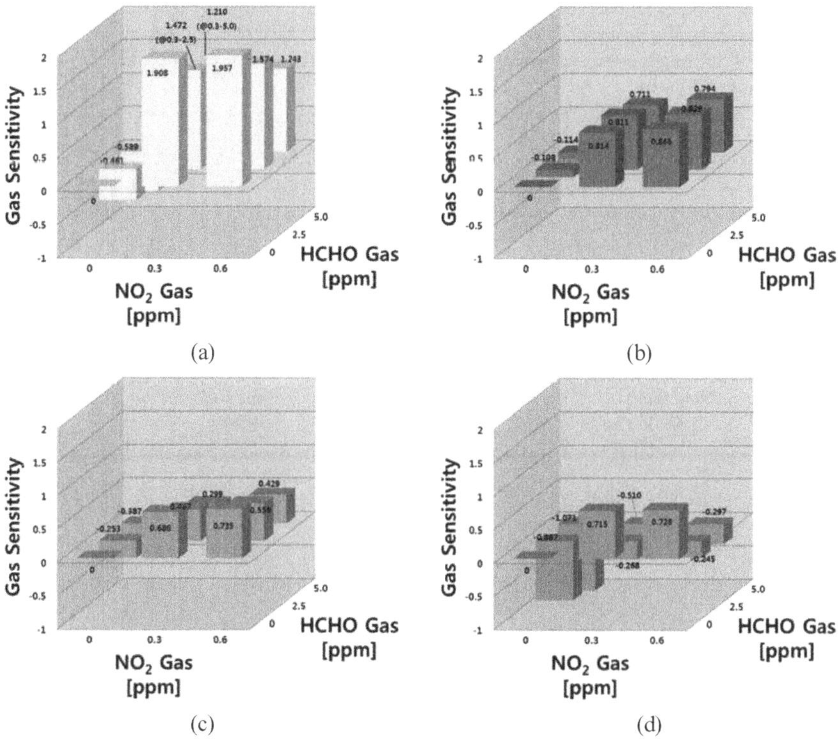

Figure 7. Gas sensing properties in the NO_2-HCHO system; (a) SN, (b) IN, (c) WO, and (d) SZ sensors.

For the final set (NH_3-HCHO system), the gas responses of the SN, WO and SZ sensors showed similar behavior in both reducing agents systems (CO-NH_3, and CO-HCHO systems), whereas the IN sensor showed no responses to both gases (in Figure 8). Three sensors (SN, WO and SZ) were sensitive to both the single gases and their mixtures, with higher sensitivities observed with the gas mixtures, and the SZ sensor responses toward HCHO were stronger than to NH_3 gas. In this system, consistent with the two reducing agents, there was no synergic effect in the SN sensor responses to a mixture of NH_3 and HCHO.

Overall, the IN sensor could detect NO_2 selectively, whereas the SN sensors detected all four (CO, NO_2, NH_3, and HCHO) gases. If coupled with an IN sensor, SN is capable of detecting NO_2 sensitively. On the other hand, the gas sensitivity signals of the two sensors were not sufficient for the detection of all four gases. The WO and SZ sensors detected all four gases but had low gas selectivity. Therefore, the four-sensor-array would be sufficient to discriminate mixtures of these gases. To gain clear insight into the applicability of the sensor array in this application, the responses with sensitivity can be arranged in a 4×4 matrix, in which each element represents the response of each sensor to each target gas. This matrix suggests how the different gas contributions can be extrapolated from the sensor array data using the signal process.

Recently, many research groups have examined the sensor response to various harmful gases and their mixtures, and reported the gas sensing patterns[21-23]. In particular, they characterized multi-sensor responses, analyzed the feature abstraction of the raw signals, and recognized the data patterns for the accurate monitoring of harmful gases. The present study analyzed the sensing properties of micro gas sensor arrays by examining the raw data in the gas mixtures.

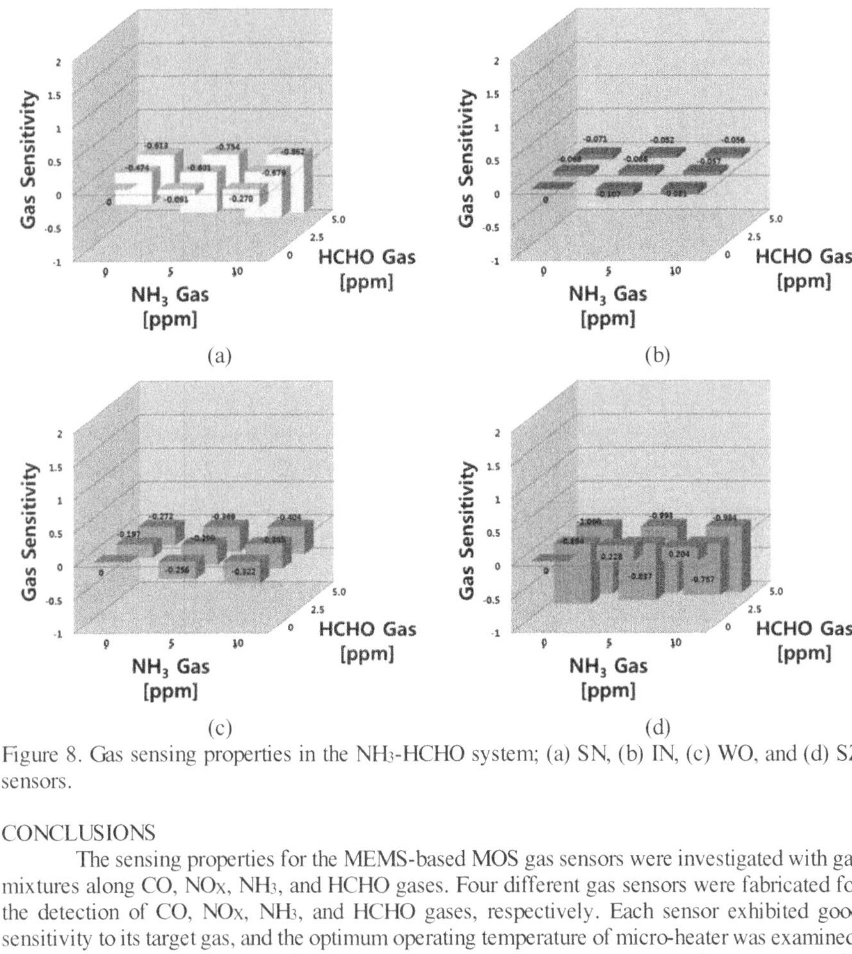

Figure 8. Gas sensing properties in the NH_3-HCHO system; (a) SN, (b) IN, (c) WO, and (d) SZ sensors.

CONCLUSIONS

The sensing properties for the MEMS-based MOS gas sensors were investigated with gas mixtures along CO, NO_X, NH_3, and HCHO gases. Four different gas sensors were fabricated for the detection of CO, NO_X, NH_3, and HCHO gases, respectively. Each sensor exhibited good sensitivity to its target gas, and the optimum operating temperature of micro-heater was examined. The sensing response behaviors for gas mixture were analyzed using the experimental data in the MEMS gas sensor arrays with respect to selectivity and response pattern. The gas sensing behaviors in mixed gas systems suggest that specific adsorption and selective activation of adsorption sites might occur in gas mixtures and offer priority for the adsorption of a specific gas. Thus, the sensor-array can discriminate a certain gas along the gas mixture with reliable selectivity. In the case of this study, the In_2O_3 based sensor (IN) detects NO_2 gas well along any mixtures of NO_2 with CO, NH_3 and HCHO. An analysis of the sensing performance of the sensor arrays will make it possible to discriminate the components in harmful gas mixtures as well as their concentrations using pattern recognition techniques.

ACKNOWLEDGEMENTS
This work was supported by the National Research Foundation of Korea (NRF) funded by the Korea government (MSIP) (No. 2015-R1A2A2A01005790).

REFERENCES

[1] Barsan, N., Koziej, D., & Weimar, U. (2008). Metal Oxide-Based Gas Sensor Research: How to. *Sensors and Actuators B: Chemical*, Volume 128, 586-593.
[2] Korotcenkov, G. (2005). Gas Response Control Through Structural and Chemical Modification of Metal Oxide Films: State of the Art and Approaches. *Sensors and Actuators B: Chemical, Volume 107*, 209-232.
[3] Vergara, A., Llobet, E., Brezmes, J., Ivanov, P., Cane, C., Gracia, I., Vilanova, X., & Correig, X. (2007). Quantitative Gas Mixture Analysis Using Temperature-modulated Micro-hotplate Gas Sensors: Selection and Validation of the Optimal Modulating Frequencies. *Sensors and Actuator B: Chemical, Volume 123*, 1002-1016.
[4] Kiselev, I., Sommer, M., Mann, J. K., & Sysoev, V.V. (2010). Employment of Electric Potential to Build a Gas Selectivity Response of Metal Oxide Gas Sensor Array. *IEEE Sensors Journal, Volume 10*, 849-855.
[5] Persaud, K., & Dodd, G. (1982). Analysis of Discrimination Mechanisms in the Mammalian Olfactory System Using a Model Nose. *Nature, Volume 299*, 352-355.
[6] Weimar, U., & Gopel, W. (1998). Chemical Imaging. Part II: Trends in Practical Multiparameter Sensors Systems. *Sensors and Actuators B: Chemical, Volume 52*, 143-161.
[7] Wang, C., Yin, L., Zhang, L., Xiang, D., & Gao, R. (2010). Metal Oxide Gas Sensor: Sensitivity and Influencing Factors. *Sensors, Volume 10*, 2088-2106.
[8] Eranna, G, Joshi, B. C., Runthala, D. P., & Gupta, R. P. (2004). Oxide Materials for Development of Integrated Gas Sensors — A Comprehensive Review. *Critical Reviews in Solid State Materials Sciences, Volume 29*, 111-188.
[9] Hinz, P., Goschnick, J., Ehrmann, S., & Ache, H. J. (1996). Multisensor Mcrosystem for Contaminants in Air. *Sensors and Actuators B: Chemical*, Volume 33, 72-76.
[10] Ivanov, P., Llobet, E., Vilanova, E., Brezmes, J., Hubalek, J., & Correig, X. (2004). Development of High Sensitivity Ethanol Gas Sensors Based on Pt-Doped SnO_2 Surfaces. *Sensors and Actuators B: Chemical, Volume 99*, 201-206.
[11] Sysoev, V. V., Schneider, T., Goschnick, J., Kiselev, I., Strelcov, E., & Kolmakov A. (2009). Percolating SnO_2 Nanowire Network as a Stable Gas Sensor: Direct Comparison of Long-Term Performance Versus SnO_2 Nanoparticle Films. *Sensors and Actuators B: Chemical, Volume 139*, 699-703.
[12] Sysoev, V. V., Button, K. B., Wepsiec, K., Dmitriev, S., & Kolmakov, A. (2006) Toward the Nanoscopic "Electronic Nose": Hydrogen vs Carbon Monoxide Discrimination with an Array of Individual Metal Oxide Nano and Mesowire Sensors. *Nano Letters, Volume 6*, 1584-1588.
[13] Huang, H., Lee, Y. C., Chow, C. L., Tan, O. K., Tse, M. S., & White, T. (2009). Plasma Treatment of SnO_2 Nanocolumn Arrays Deposited by Liquid Injection Plasma-Enhanced Chemical Vapor Deposition for Gas Sensors. *Sensors and Actuators B: Chemical, Volume 138*, 201-206.
[14] Minghua, S. (2006) Gas Identification Based on Committee Machine for Microelectronic Gas Sensor, *IEEE Transactions on Instrument Measurement, Volume 55*, 1786-1793.
[15] Zhang, S., Xie, C., Zeng, D., Li, H., Bai, Z., & Cai, S. (2008). A Method of Feature Extraction from the Desorption Part of MOX's Response Curves to Gases. *IEEE Sensors Journal, Volume 8*,

1816-1823.

[16] Kim, S. D., Kim, B. J., Yoon. J. H., & Kim, J. S. (2007). Design, Fabrication and Characterization of a Low-Power Gas Sensor with High Sensitivity to CO Gas. *Journal of Korean Physics Society, Volume 51*, 2069-2076.

[17] Yoon, J. H., & Kim, J. S. (2009) Design and Fabrication of a MEMS-Based Gas Sensor. *Advanced Material Research, Volume 74*, 255-258.

[18] Lee, H. J., Yoon, J. H., Kim, B. J., Jang, H. D., & Kim, J. S. (2010). Gas Sensing Characteristics of Ru Doped-WO_3 Micro Gas Sensors. *Korean Journal of Metals and Materials, Volume 49*, 395-399.

[19] Kim, B. J., Lee, H. J., Yoon, J. H., & Kim, J. S. (2012) Highly Sensitive Formaldehyde Gas Sensors Based on SnO_2-ZnO Nanocomposites. *Sensors Letters, Volume 10*, 1-7.

[20] Choi, W. S., Kim, B. J., Lee, H. J., Choi, J. W., Kim, S. D., & Min, N. K. (2012) Study on the Micro-Heater Geometry in MEMS Gas Sensor Platforms and Effects on Gas Detecting Performances. *Journal of Nanoscience and Nanotechnology, Volume 12*, 1-4.

[21] Liao, F., Yin, S., Toney, M. F., & Subramanian, V. (2010). Physical Discrimination of Amine Vapor Mixtures Using Polythiophene Gas Sensor Arrays. *Sensors and Actuators B: Chemical, Volume 150*, 254-263.

[22] Zhao, D., Zhang, Y., Kong, D., Chen, Q., & Lin, H. (2012). Research on Recognition System of Agriculture Products Gas Sensor Array and its Application, *Procedia Engineering*, Volume 29, 2252-2256.

[23] Barbri, N. E., Mirhisse, J., Ionescu, R., Bari, N. E., Correig, X., Bouchikhi, B., & Llobet, E. (2009). An electronic nose system based on a micro-machined gas sensor array to assess the freshness of sardines. *Sensors and Actuators B: Chemical, Volume 141*, 538-543.

Author Index

Almazán, F., 139
Alves, A. K., 3, 13
Assi, Z., 105

Bakshi, A., 121
Bergmann, C. P., 3, 13
Berutti, F. A., 3, 13
Buhl, J. C., 105

Cai, Q., 43

Da Dalt, S., 13
Debéda, H., 139

Garcia, A. P., 13

Hayashi, H., 69
Hirata, Y., 59

Jubéra, V., 139
Jung, S.-C., 23

Kim, J.-S., 23, 147
Kim, B.-J., 147
Kishimoto, A., 69
Kuhn, M., 121

Lee, J.-W., 129
Li, H., 43
Lucat, C., 139

Moeller, M., 121

Neufert, R., 121
Nguyen, V. S., 139

Ojima, H., 35
Okuwa, T., 81
Onuki, T., 35

Pawlik, M., 93
Pernechele, M., 93
Pilar, M. P., 139

Rodrigues, F., 121
Rüscher, C. H., 105

Sameshima, S., 59
Schomborg, L., 105
Sheridan, E., 121
Shimizu, J., 35
Shimonosono, T., 59
Sung, H.-J., 23

Teloeken, A. C., 3
Teranishi, T., 69
Tohji, A., 69
Troczynski, T., 93

Ueno, S., 129

Vincent, A., 121

Wada, H., 81
Wang, J., 43

Wu, J., 43

Yahiro, H., 81
Yamaguchi, S., 81
Yamamoto, T., 35

Yang, Y., 43

Zhao, B., 43
Zhou, L., 35

Lightning Source UK Ltd.
Milton Keynes UK
UKOW06n1432150816

280740UK00001B/96/P

9 781119 234449